The Physiology of Aerobic Capacity in Women

This book questions the limitation of exercise capacity in women by discussing female physiology from the perspectives of respiratory, circulatory, skeletal, body composition, and training adaptations. Written in a compelling manner, the book covers not only gender differences in exercise physiology but also touches upon such questions as doping and novel mechanisms in exercise theory and practice. Based on first-hand research experience, this book offers new and realistic perspectives, including positive and negative aspects of women's capacity to perform exercise, which should interest the readers of kinesiology, integrative physiology, clinical science, general science and sociology of sports topics.

Key Features:

- Research-based findings on the cutting-edge topic of women's aerobic capacity.
- Written in an accessible manner and packed with science-based insights.
- Presents an overarching view of various medical disciplines that are essential in evaluating women's aerobic capacity.

The Physiology of Aerobic Capacity in Women

David Montero Barril

CRC Press is an imprint of the
Taylor & Francis Group, an **informa** business

Designed cover image: © Stock Photo ID: 254204632, Photo Contributor Izf

First edition published 2025
by CRC Press
2385 NW Executive Center Drive, Suite 320, Boca Raton FL 33431

and by CRC Press
4 Park Square, Milton Park, Abingdon, Oxon, OX14 4RN

CRC Press is an imprint of Taylor & Francis Group, LLC

ISBN: 978-1-032-78231-7 (hbk)
ISBN: 978-1-032-77693-4 (pbk)
ISBN: 978-1-003-48689-3 (ebk)

DOI: 10.1201/9781003486893

Typeset in Sabon
by SPi Technologies India Pvt Ltd (Straive)

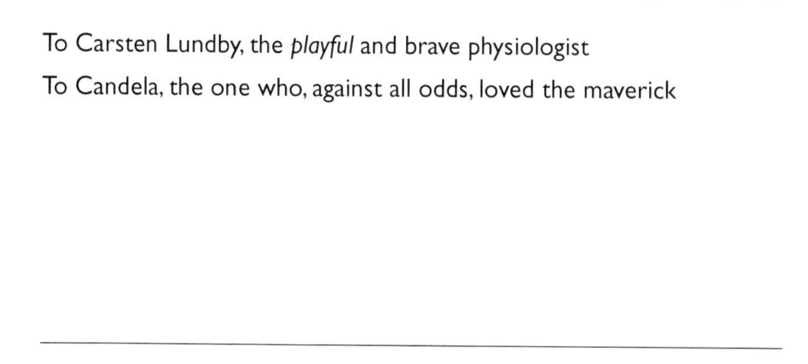

To Carsten Lundby, the *playful* and brave physiologist

To Candela, the one who, against all odds, loved the maverick

Contents

Preface

Why do women generally display lower exercise capacity than men? Is this due to innate biological characteristics? Notwithstanding the general set of instructions that we all carry in the genetic code, can women adapt so as to reach or even surpass men's performance? Many will hold at least intuitive answers to these questions. A minority, likely scholarly related individuals, will be confident to know. How much do we really know? We should never cease to regularly ask ourselves this question, among others. Questions—the right ones—move us forward. Answers, right or wrong, lead to complacency. When the question posed is somehow expected, i.e., follows the mainstream, wrong answers are highly prevalent. It is more likely to be propelled forward (into truer knowledge) by questions that either have never been—and seem too forthright to be—asked, or whose answers are supposed to be known from a long time ago, decades or centuries.

This book gradually emerged as a result of a myriad of 'forward-propelling' questions about how the human body works, along with some answers. No wonder, that the mere fact of questioning the scholar *status quo* of our time required struggle and stamina. Above all, it required one to take a substantial risk of virtually ending an incipient academic career. An uncompromising will to search for and present the most genuine findings is indeed a suicidal trait among scholars, notably for 'rookies'. Yet, is there any worthwhile endeavor that does not demand some kind of quixotism? Even serendipity requires the will to wander, and implicitly a faith in wandering.

Let us not confound the harshness of the journey with the outcome. The relevance of any scientific work must be valued independently of the external circumstances in which it was developed. In the scientific arena, we should always hold our emotional judgement in check. Indeed, the scientific method is a means to prevent being *too* human. The better the methodology, in the broader sense of the term, the greater the potential relevance regardless of the personal attributes and scientific 'purity' of the investigators. Accordingly, critical discoveries are more likely to be made in a mature scientific field with access to advanced methods emerging from the inexorable technological progress in human history. In this regard, we may admire the sagacity and perceptiveness, relative to ancient available methods, but not the pertinence of Aristotle's physiological writings. No matter the emphasis and gravity of expression, the facts, interpretations, hypotheses and conclusions presented in this book must be weighted, without exception, conforming to the quality of the underlying methods. This is an effort made, albeit its intricacies are seldom disclosed due to the divulgatory nature of this work, by the author.

As counterintuitive as it may seem, research into the inner workings of our body entails, to some extent, subjectivism and senselessness. The researchers must decide the research question, the means to address it (i.e., the methodology, including a frequently ill-prepared student's workforce), as well as the presentation and interpretation of results. At each step multiple choices are made, with generally no available guideline or established recommendation to follow. A seemingly trivial change in the selection of methods, e.g., among measuring instruments or statistical tests, may result in significantly distinct or even opposite findings. Moreover, investigations using humans as experimental subjects encounter plenty of constraints, mainly of an ethical nature, with a varied degree of rational justification or, rather, based on absolute ignorance. The possibility of using animals for research, typically rodents, is also limited, as for many integrative questions their quadrupedal physiology is distinct enough to accurately extrapolate to humans. In order to unequivocally elucidate the physiology of exercise capacity and its sex-specific traits, we need to experiment in women and men.

Humans are endowed with a wide spectrum of partly independent exercise capacities. Single heavy lifts lasting seconds, repeated power outbursts for a few minutes, endurance efforts sustained for hours, and anything in between, require a diverse predominant contribution of specific metabolic pathways and varying involvement of neuromuscular and cardiovascular systems. Our body cannot concurrently have an all-embracing outstanding exercise capacity. The reason is twofold. First, in order to excel in a definite exercise performance, humans need at least months, frequently years, of exercise training precisely targeting—i.e., challenging—the physiological systems involved in that performance. These training stimuli progressively induce adaptations that enhance the corresponding, but not necessarily other, exercise capacity. Second, training adaptations may be complementary, but some are antagonistic. For instance, large gains in muscle mass induced by resistance training improve strength but impair prolonged weight-bearing exercise such as long-distance running or cycling uphill—provided that before starting the exercise training program muscle mass is not low enough to limit endurance performance. Conversely, neuromuscular adaptations to endurance training (ET), while enhancing the recruitment of fatigue-resistant muscle fibers, restrain muscular strength and power. Taken together, the term 'exercise capacity' is very unspecific. We must choose a definite exercise capacity to unequivocally answer the opening questions. Alternatively, we can focus on a physiological variable underpinning a wide spectrum of exercise capacities.

Aerobic capacity refers to the maximal amount of oxygen (O_2) that an individual can consume per unit of time (VO_{2max}). It reflects the maximal rate of energy expenditure through aerobic metabolism, which is by far the predominant supplier of energy substrates (>90%) in resting humans. Likewise, aerobic metabolism is the main energy supplier in the vast majority of exercises that a human can perform: human life is thus strictly and overwhelmingly dependent on O_2. Only in exercise bouts performed at the highest possible exercise intensity and lasting less than ~2 minutes, the aerobic contribution to energy production does not prevail. When these short (<2 minutes) maximal efforts are repeated (i.e., interspersed with periods of lower exercise intensity or rest) aerobic metabolism is fundamental to partly recovering the energy stores, therefore attenuating the decline in energy output of additional maximal efforts. VO_{2max} strongly determines exercise capacities that continuously require high aerobic metabolism (>70% VO_{2max}) for prolonged time (>10 minutes), such as those underlying the endurance performance in long-distance running, cycling, cross-country skiing, etc., collectively referred as 'aerobic exercise capacity'. Beyond determining endurance performance, VO_{2max} is one of the strongest predictors of morbidity, mortality and life expectancy. In fact, endurance athletes, whose training stimuli, over years, leads to the highest VO_{2max} in humans, can be considered as the 'healthiest' specimens of our species. It is no wonder that more is known about VO_{2max} than for the rest of physiological variables underlying exercise capacities together. Hence, this book focuses on VO_{2max} and its physiological determinants.

The author

David Montero Barril is jointly appointed by the Department of Medicine, School of Clinical Medicine, and the School of Public Health at Hong Kong University. Following a PhD project funded by the French Society of Vascular Medicine, he received postdoctoral training at the Cardiovascular Research Institute of Maastricht (Netherlands), the Zürich Center for Integrative Human Physiology and the Department of Cardiology of the University Hospital of Zürich (Switzerland). Prof. Montero welcomes challenging questions, specifically those with the potential to arise the necessary intrinsic motivation to be enthusiastically immersed in them. Current research questions converge upon essential mechanisms underpinning the cardiovascular capacity to deliver oxygen to the tissues, one of the strongest performance (endurance) and clinical (all-cause mortality) predictors. Embracing integrative approaches, his laboratory focuses on the controlled manipulation and accurate acquisition of the interplay between cardiovascular, hematological, nervous and metabolic systems during physiologically relevant conditions. Exercise is implemented as a means to magnify and thereby facilitate the understanding of key intertwined mechanisms of the human body in health and disease.

Chapter 1

Respiratory system

Abbreviations:

EIAH, exercise-induced arterial hypoxemia
CO_2, carbon dioxide
O_2, oxygen
VO_{2max}, maximal oxygen consumption

ASSEMBLY LINES AND BOTTLENECKS IN BIOLOGICAL DELIVERY SYSTEMS

The respiratory system encompasses an airway tract that links the external environment with our internal circulatory system. Anatomically, from outside to inside, the airway is comprised of: the nose, nasal cavity, pharynx, larynx, trachea, bronchi, bronchioles and alveoli in the lungs. O_2 follows this route to diffuse into the blood, the fluid vehicle that delivers the gas (O_2) into the tissues. For any linear process involving step-by-step stages, as, for instance, the typical assembly line in the automobile industry, there must be a rate-limiting step, i.e., a slowest step. The maximal speed of the assembly line can be only as fast as that of the slowest step. This might be the engine's installation step, the painting step, the drying step, etc. A single factor will set the rhythm. In theory, all steps could have exactly the same speed, hence there would be no single weakest link. In reality, such a harmonious concert is highly unlikely in industrial manufacturing.

A linear process without a weakest link is still more improbable in biology, in which irregularities and inherent variability are the norm. Specifically, the serial constituents of the O_2 transport chain from the atmosphere into the mitochondria in skeletal muscle

DOI: 10.1201/9781003486893-1

fibers, where most O_2 is finally consumed during exercise, are far from being perfectly tuned with each other. Namely, the capacity to transport O_2 from the respiratory to the circulatory system is much faster than that from the heart to the mitochondria in skeletal muscle fibers. In other words, the O_2 transport chain is *asymmetrical*; not all steps have the same transport speed and thus the same level of importance in limiting maximal oxygen consumption (VO_{2max}).[1]

In a minute, the respiratory system of a healthy sedentary man can *voluntarily* move in up to ~150 L of air, i.e., ~31 L of O_2.[2] This number reflects the total volume of O_2 entering the airway tract, which must be greater than the O_2 reaching the alveoli, where O_2 diffuses into the circulation. A fraction of the inspired air remains in regions of the airway tract that do not contribute to gas exchange (from the nose up to the bronchi), known as the anatomical dead space. Approximately a third of the gas in every inspiration never reaches the gas exchange region. The same applies for every expiration: a third of its content, comprising inert gases, water vapor, unabsorbed O_2 as well as a fraction of CO_2 diffusing from the circulation, do not exit the airway tract. In this regard, the CO_2 generated by metabolic processes in the body and diffusing into the alveoli adds to the water vapor that fully saturates the airway tract, both CO_2 and water vapor contributing to reduce the available space for other gases, including O_2. Considering the anatomical dead space and the typical 'amount' of CO_2 and water vapor in the alveoli (as reflected by their partial pressure, as will be explained), only ~19 L of O_2 can reach the gas exchange region of our sedentary man at maximal voluntary ventilation. In striking contrast, this individual may deliver no more than ~3.5 L of O_2 to the tissues via the circulatory system at maximal effort. Thus, the bottleneck—i.e., the weakest link—cannot be in the respiratory system.

The diffusion of O_2, as well as other gases, between the alveoli and the pulmonary capillaries—i.e., the interface between the respiratory and circulatory systems—is a passive process. Energy is not required at the molecular level to propel O_2 into the circulation. It happens via a concentration gradient. When referring to gases, the usual term to convey the 'concentration' is partial pressure, defined as the pressure exerted by an individual gas in a mixture of gases. The concentration of a gas is directly proportional to its partial pressure under stable conditions. Therefore, the partial pressure of O_2 in the alveoli must be higher than that in pulmonary capillary blood in order that O_2 diffuses into the circulation. Such a gradient is always present to some extent, given that blood O_2 is relentlessly consumed and not produced in the living human body. O_2 thus continuously flows into the circulation, where it will encounter greater transport limitations, as we will develop in the following chapters.

By now, the question arises as to whether the respiratory system plays a role in sex differences in aerobic exercise capacity. As aforementioned, provided that healthy individuals can deliver a fivefold excess of O_2 from the external environment to the internal carrier (the circulation), no potential limitation of VO_{2max} can seemingly be imputed to the respiratory system. Yet, sex differences in respiratory efficiency as well as external conditions could have a relevant impact.

SEX DIFFERENCES IN THE RESPIRATORY SYSTEM: DO THEY MATTER FOR AEROBIC EXERCISE CAPACITY?

Divergences in size typically pervade the discussion of sex differences. The prevalently lower body size of women is necessarily reflected in major structures of the respiratory system (Figure 1.1). For instance, the rib cage and shape of the lungs consistently differ between sexes, with women having reduced base and overall size of the lungs.[3] Yet, the respiratory system of women is also smaller compared with men when controlling for anthropometric differences.[4] Even when adjusted by lung size, the cross-sectional area of large conducting airways (trachea and bronchi) is reduced by up to one third in women relative to men, which entails substantial functional repercussions.[4] According to the physical law describing laminar flow—developed in detail in the following chapter focusing on the circulatory system—the resistance to flow is inversely proportional to the radius of the cross-sectional area of the airway to the fourth power. Therefore, the metabolic work of breathing must be *constitutionally* higher in women.

In resting conditions, sex differences in the respiratory system do not have a meaningful impact on respiratory function. When the O_2 demand and ventilation increases, e.g., with exercise, the magnitude of the effect of sex is evident. In this regard, at any given level of ventilation above 55 L/min—required in

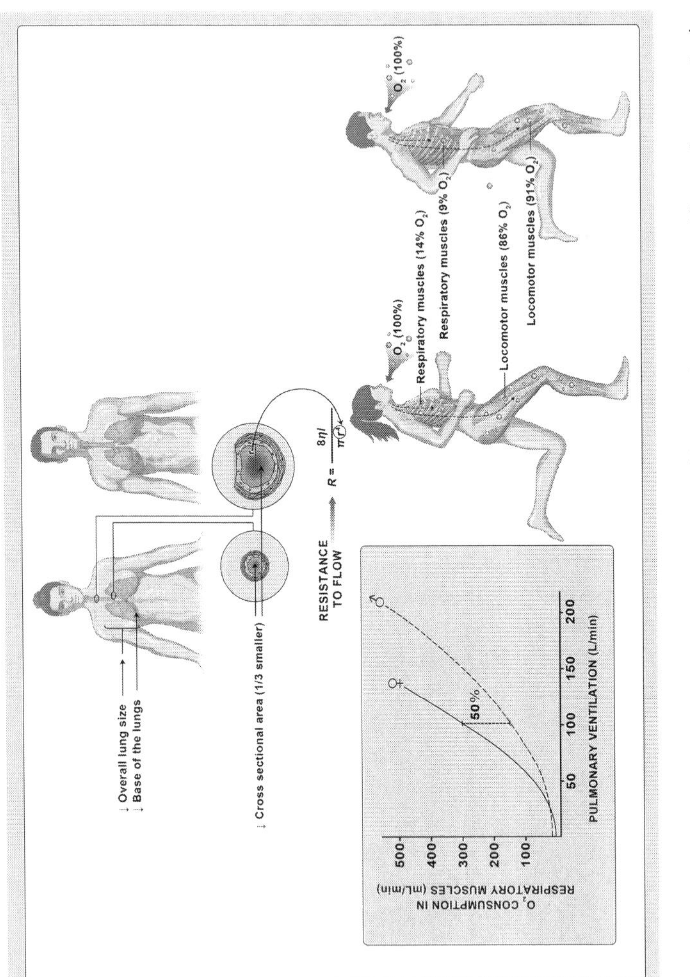

Figure 1.1 Main sex differences in the respiratory system: Functional consequences. Men and women in the figure have the same body weight. *l*, length of the conduit; *n*, viscosity of the fluid; *R*, resistance to flow; r, radius; *π*, pi.

moderate-to-high intensity exercise—the respiratory work and thus the metabolic cost of breathing (represented by the O_2 consumption of respiratory muscles) is exponentially augmented in women compared with men.[5] Remarkably, the cost of breathing at very high ventilation levels (>100 L/min) is ≥50% higher in women relative to men.[5, 6] Such a sex dimorphism is observed independent of fitness status and body size and is accentuated with older age.[5, 6] As expected, owing to the smaller airways in women, the increased resistive work of breathing is the main underlying contributing factor to the higher metabolic cost.[6] Moreover, such increased work requires higher positive intrathoracic pressures on expiration, which may decrease cardiac filling and therefore impair cardiac pumping capacity and circulatory O_2 delivery, conforming to the Law of the Heart (detailed in the following chapter).[7] Limited structure results in less effective interaction among body systems.

The preceding discussion has implications for VO_{2max}. At the upper end of the aerobic capacity spectrum, VO_{2max} reaches ~85 and ~72 mL/min per kg of body weight in men and women elite endurance athletes, respectively.[8] No more O_2 can be absorbed by the human body. Hence, the extra O_2 that the respiratory muscles of women consume for a given level ventilation during exercise is not available to the active locomotive muscles.[9] Women's respiratory muscles at peak exercise consume 14% of VO_{2max}, while that of men consume only 9%.[10] Consequently, the aerobic potential for force production (contraction) in the locomotive muscles of women is proportionally reduced. In the hypothetical scenario that no further sex differences in the O_2 transport and utilization chain were present, women would be less efficient during exercise than men. For every mL of O_2 consumed (a surrogate of energy expenditure), the exercise performance (e.g., power output on a bicycle or running speed) would be lower in women compared with men matched by body weight. Nonetheless, sex differences in exercise efficiency have not been conclusively established.[11] In fact, recent investigations in our laboratory comprising a large sample size (>200) of healthy women and men reveal higher exercise efficiency in women.[27] It is plausible that downstream factors in the O_2 cascade (developed in Chapters 4 and 5) differ between sexes, thereby counterbalancing the female respiratory *handicap* during exercise.

FAILURE OF THE RESPIRATORY SYSTEM AT WORK: EXERCISE-INDUCED ARTERIAL HYPOXEMIA

If human physiology operated with the presumed precision of a Swiss watch, the respiratory system would deliver to the circulatory system the maximum amount of O_2 that can be carried by the blood. Such a level of O_2 is fundamentally determined by the protein carrier in blood, i.e., hemoglobin. The more hemoglobin in blood, the higher the O_2 carrying capacity. In this ideal scenario, blood, specifically the hemoglobin in arterial blood, will be fully saturated (100%) with O_2. In practice, however, arterial blood is seldom fully saturated with O_2. Normal values commonly range from 95% to 99%. It is worth mentioning that the term 'normal' in biological sciences is not as definite as we make it frequently appear.

In resting conditions, blood O_2 saturation below 90% is termed hypoxemia. A disease or substantial alteration of the respiratory system must underlie hypoxemia at rest. In contrast, seemingly healthy individuals may exhibit low O_2 saturation during exercise, which is not a minor issue for exercise performance given that saturated arterial blood is essential to maximize O_2 delivery to the systemic circulation, a fundamental determinant of VO_{2max}.[12] The so-called exercise-induced arterial hypoxemia (EIAH) is defined as a decrease in arterial oxygen saturation below 94%.[13] As expected, decrements in arterial O_2 content, VO_{2max} and endurance performance (e.g., time trial) are strongly related.[14, 15] The prevalence of EIAH in the general population is still unclear due the absence of large population studies. Available evidence estimates the prevalence of EIAH between 20% and 50%.[16] Yet, experts in the field remain skeptical about the validity of such high estimates.[17]

What has been mainly established thus far is the increasing likelihood of EIAH with higher fitness (VO_{2max}). The higher the VO_{2max}, the greater the cardiac capacity to pump blood to the body, including the lungs. From these facts it is speculated that highly fit individuals have less 'reserve' in the respiratory system to saturate with O_2 the very high blood flow perfusing the alveoli. Of note, cardiovascular and metabolic systems readily respond (adapt) to exercise training, while the respiratory system is mainly unresponsive.[13] Hence, the enhanced cardiovascular and metabolic capacity of aerobically fit individuals is more likely to reach the upper limit of the respiratory system, according to the aforementioned construct.

As we will develop further in the next section, the respiratory system can be challenged to adapt in specific limiting conditions, such as at very high altitude. At sea level or moderate altitude, respiratory structure and function is 'overbuilt' in relation to the maximally achieved cardiovascular and metabolic demands of healthy sedentary to moderately trained individuals.[13]

For any given fitness level, women exhibit higher prevalence of EIAH than men in the majority of studies available to date.[16, 17] Even women with low VO_{2max} (i.e., sedentary) can develop EIAH, which is seldom observed in healthy sedentary men.[18] This can be explained, at least in part, by the aforementioned female-specific respiratory constraints.[5, 6] Due to the smaller airway cross-sectional area, women are prone to mechanical ventilatory limitations leading to alveolar hypoventilation from moderate to peak exercise.[18] Remarkably, when the work of breathing is experimentally reduced via the inclusion of heliox (a low-density gas) in the breathing mixture, EIAH is reversed in women exhibiting large respiratory constraints.[18] In summary, in addition to the inefficiency in the cost of breathing and possibly altered cardiac mechanics primarily caused by reduced airways, women are more likely to suffer from reduced O_2 delivery attributed to blood O_2 desaturation during exercise. This is not a trivial set of sex-specific constraints to start with the O_2 cascade, from the air into the mitochondria.

ALTITUDE AND O_2 AVAILABILITY

When describing human physiology, among other assumptions, we tacitly refer to individuals at sea level, therefore with plenty of O_2 available. While the percentage of O_2 in the atmosphere is similar regardless of the location, the pressure, which represents the total amount of gas molecules bouncing around in a given space, gradually decreases with altitude. At sea level, the pressure of the column of air above us is ~760 mm of mercury (mm Hg), known as 1 atmosphere. Translated to weight units, each cm^2 of our body surface is subjected to the pressure of 1.03 kg. In total, the body of an average man with a typical body surface area of 1.9 m^2 is subjected to more than 19,000 kg at sea level—we do not notice. Without such a pressure, i.e., in the absence of air (no pressure), such as in outer space, the unprotected human body would partly disintegrate. In fact, in the majestic interplanetary void, the gas inside the respiratory system would immediately expand, causing the lungs to rupture if air is not exhaled promptly. Concomitantly, the liquid water in the body would turn into water vapor, with the internal membranes and skin swelling like an inflated balloon. We are lucky if we do not explode.

Coming closer to the earth, the lack of air is noticed at altitudes above sea level in that human life and thus athletic events are feasible. From sea level to 3000 m of altitude, the atmospheric pressure is reduced by ~80 mm Hg (11% decrement) for every 1000 m gain. Above 3000 m, the drop in pressure is partly attenuated, given that approximately 40% of all air molecules are located in the lower 4000 m, a minimal fraction of the atmosphere's height (>100 km). The respiratory muscles at altitude must work more so that the alveoli *offer* to the circulatory system the same amount of O_2 as at sea level. However, it can be questioned whether the respiratory system has to deliver that (sea level) amount of O_2. Perhaps, under normal (involuntary) respiratory conditions, there may be an excess of O_2 being offered and hence no need to increase ventilation. In other words, we can question whether humans ventilate more than is required. It might appear so through looking at the remaining O_2 in the venous blood coming back to the lungs. At rest, about 70% of O_2 in the circulatory system is not consumed. Even at maximal exercise, 10% to 20% of the O_2 in blood is left unused, albeit not all blood flow is directed towards tissues in high need of O_2 (e.g., active muscles).[19] Yet, the large surplus of O_2 at rest suggests a biological luxury in O_2 delivery, which may not be extrapolated to maximal exercise conditions, where the capacity to deliver blood and O_2 to the active muscles is fully exploited.[19] Before we jump to conclusions, let's consider the fundamental concepts in the next section.

THE REGULATION AND INCREASING IMPORTANCE OF VENTILATION WITH ALTITUDE

Ventilation is primarily regulated by the concentration in blood of CO_2, the end product of multiple metabolic pathways. CO_2 accumulates in the circulation for any degree of hypoventilation. Consequently, normal ventilation is necessary to keep a rather fixed blood CO_2 concentration. The accumulation of CO_2 in the circulation and the associated acidosis, rather than the lack of O_2, is the respiratory variable that is more stringently controlled—this is how evolution has unfolded. We know from laboratory experiments (independently manipulating the amount of O_2 and CO_2 in blood) that the absolute O_2 content in the circulation, which is mainly determined by blood O_2 carrying capacity, is not sensed by the respiratory centers in the brain that control ventilation.[20] In contrast, the partial pressure of O_2 in blood is detected, but it plays only a secondary role in the regulation of ventilation. Of note, the absolute content and partial pressure of O_2 may vastly differ due to predominant effect of the O_2 carrier in blood (the protein hemoglobin) on absolute O_2 content, as it will be developed in Chapter 3. Notwithstanding, whether humans ventilate more or less than *needed* should be primarily framed according to CO_2.

Importantly, if circulating CO_2 is not augmented, remaining at or below the basal level is determined by homeostatic mechanisms, so it should be circulating O_2 but in the opposite sense (i.e., not reduced) under most conditions, provided that the regulations of CO_2 and O_2 were coupled. Yet, already at the mild altitudes of 580 m and 750 m above sea level, blood O_2 saturation is reduced, and thus the potential O_2 delivery to active tissues is diminished, particularly during exercise.[21, 22] Consequently, the respiratory system is not able to completely fulfill its O_2 delivery function even at moderate altitudes, despite the fact that ventilation is augmented to maintain or reduce blood CO_2 concentration. Hence, if any excess of O_2 is being delivered by the respiratory to the circulatory system at sea level, it must be minor.

As it follows from reduced blood O_2 saturation, aerobic capacity is largely affected by altitude. VO_{2max} is reduced by ~6–10% for every 1000 m increase in altitude from sea level to 3000 m.[23] At the highest surface point of the earth, the top of Mount Everest (8848 m), an ~80% decrement in VO_{2max} is present relative to sea level. A very fit climber with a VO_{2max} of 75 mL/min/kg at sea level will end up with 15 mL/min/kg at the Everest summit, scarcely above the threshold for cardiac transplantation (12 mL/min/kg). With such an aerobic capacity, even slow walking is a challenge for the fittest humans in the absence of extra (artificial) O_2 support. Taken together, at high altitude, the respiratory system is a limiting factor, a bottleneck, for VO_{2max}.

IS WOMEN'S EXERCISE CAPACITY MORE AFFECTED BY ALTITUDE?

Provided that (1) the female respiratory system has a limited structural and functional reserve compared with men, and (2) high altitude necessarily challenges such a reserve by increasing the respiratory work for a given absolute workload, the answer might be yes, notwithstanding that the available evidence is scarce. Sex comparisons regarding the decrement in aerobic exercise performance remain to be systematically investigated. Limited data suggest that the high altitude (>1500 m)-induced impairment of endurance performance in long-distance running (10,000 m) is greater in elite female endurance athletes than in male counterparts.[24] Even if the previous findings were confirmed and could be extrapolated to other exercise modalities and fitness status, plenty of questions would still be unanswered. For instance, athletes commonly spend several days to weeks acclimatizing at high altitude when the competition is performed in high altitude; do women need more time to acclimatize than men? Again, the available evidence is indirect and still inconclusive.[25] Looking into non-exercise findings, the frequent symptoms (headache, anorexia, nausea, vomiting, dyspnea, lassitude, insomnia) that individuals living at sea level experience when they ascent to high altitude (>2500 m) are 24% more prevalent in women, regardless of age or race.[26] Yet, the sex-specific impact of altitude on major physiological adaptations determining exercise, particularly endurance performance, will have to substantiated in future studies. In science, answers are contingent on data.

REFERENCES

1. Lundby C, Montero D. CrossTalk opposing view: Diffusion limitation of O_2 from microvessels into muscle does not contribute to the limitation of VO2 max. *J Physiol.* 2015;593(17):3759–3761.
2. Otto-Yanez M, Sarmento da Nobrega AJ, Torres-Castro R, et al. Maximal voluntary ventilation should not be estimated from the forced expiratory volume in the first second in healthy people and COPD patients. *Front Physiol.* 2020;11:537.
3. Torres-Tamayo N, Garcia-Martinez D, Lois Zlolniski S, Torres-Sanchez I, Garcia-Rio F, Bastir M. 3D analysis of sexual dimorphism

in size, shape and breathing kinematics of human lungs. *J Anat.* 2018;232(2):227–237.

4. Sheel AW, Guenette JA, Yuan R, et al. Evidence for dysanapsis using computed tomographic imaging of the airways in older ex-smokers. *J Appl Physiol.* 2009;107(5):1622–1628.
5. Molgat-Seon Y, Dominelli PB, Ramsook AH, et al. The effects of age and sex on mechanical ventilatory constraint and dyspnea during exercise in healthy humans. *J Appl Physiol.* 2018;124(4):1092–1106.
6. Dominelli PB, Molgat-Seon Y, Bingham D, et al. Dysanapsis and the resistive work of breathing during exercise in healthy men and women. *J Appl Physiol.* 2015;119(10):1105–1113.
7. Miller JD, Hemauer SJ, Smith CA, Stickland MK, Dempsey JA. Expiratory threshold loading impairs cardiovascular function in health and chronic heart failure during submaximal exercise. *J Appl.* 2006;101(1):213–227.
8. Lundby C, Robach P. Performance enhancement: What are the physiological limits? *Physiology (Bethesda).* 2015;30(4):282–292.
9. Dominelli PB, Archiza B, Ramsook AH, et al. Effects of respiratory muscle work on respiratory and locomotor blood flow during exercise. *Exp Physiol.* 2017;102(11):1535–1547.
10. Dominelli PB, Render JN, Molgat-Seon Y, Foster GE, Romer LM, Sheel AW. Oxygen cost of exercise hyperpnoea is greater in women compared with men. *J Physiol.* 2015;593(8):1965–1979.
11. Lundby C, Montero D, Gehrig S, et al. Physiological, biochemical, anthropometric, and biomechanical influences on exercise economy in humans. *Scand J Med Sci Sports.* 2017;27(12):1627–1637.
12. Lundby C, Montero D, Joyner M. Biology of VO2 max: Looking under the physiology lamp. *Acta Physiol (Oxf).* 2017;220(2):218–228.
13. Dempsey JA, La Gerche A, Hull JH. Is the healthy respiratory system built just right, overbuilt, or underbuilt to meet the demands imposed by exercise? *J Appl Physiol.* 2020;129(6):1235–1256.
14. Amann M, Eldridge MW, Lovering AT, Stickland MK, Pegelow DF, Dempsey JA. Arterial oxygenation influences central motor output and exercise performance via effects on peripheral locomotor muscle fatigue in humans. *J Physiol.* 2006;575(Pt 3):937–952.
15. Diaz-Canestro C, Siebenmann C, Montero D. Blood oxygen carrying capacity determines cardiorespiratory fitness in middle-age and older women and men. *Med Sci Sports Exerc.* 2021;53(11):2274–2282.
16. Guenette JA, Sheel AW. Exercise-induced arterial hypoxaemia in active young women. *Appl Physiol Nutr Metab.* 2007;32(6):1263–1273.
17. Dominelli PB, Sheel AW. Exercise-induced arterial hypoxemia; some answers, more questions. *Appl Physiol Nutr Metab.* 2019;44(6):571–579.

18. Dominelli PB, Foster GE, Dominelli GS, et al. Exercise-induced arterial hypoxaemia and the mechanics of breathing in healthy young women. *J Physiol.* 2013;591(12):3017–3034.
19. Skattebo O, Calbet JAL, Rud B, Capelli C, Hallen J. Contribution of oxygen extraction fraction to maximal oxygen uptake in healthy young men. *Acta Physiol (Oxf).* 2020;230(2):e13486.
20. Whitwam JG, Duffin J, Triscott A, Lewin K. Stimulation of the peripheral chemoreceptors with sodium bicarbonate. *Br J Anaesth.* 1976;48(9):853–857.
21. Goldberg S, Buhbut E, Mimouni FB, Joseph L, Picard E. Effect of moderate elevation above sea level on blood oxygen saturation in healthy young adults. *Respiration.* 2012;84(3):207–211.
22. Gore CJ, Little SC, Hahn AG, et al. Reduced performance of male and female athletes at 580 m altitude. *Eur J Appl Physiol Occup Physiol.* 1997;75(2):136–143.
23. Chapman RF, Stickford JL, Levine BD. Altitude training considerations for the winter sport athlete. *Exp Physiol.* 2010;95(3):411–421.
24. Hamlin MJ, Hopkins WG, Hollings SC. Effects of altitude on performance of elite track-and-field athletes. *Int J Sports Physiol Perform.* 2015;10(7):881–887.
25. Muza SR, Rock PB, Fulco CS, et al. Women at altitude: Ventilatory acclimatization at 4,300 m. *J Appl Physiol.* 2001;91(4):1791–1799.
26. Hou YP, Wu JL, Tan C, Chen Y, Guo R, Luo YJ. Sex-based differences in the prevalence of acute mountain sickness: A meta-analysis. *Mil Med Res.* 2019;6(1):38.
27. Guo M, Montero D. Women consume less oxygen than men per muscular work: role of lean body mass. *Mayo Clinic Proceedings.* 2024 (accepted June 2024).

Chapter 2

Circulatory system

Abbreviations:

CO_2, carbon dioxide
HR_{max}, maximal heart rate
O_2, oxygen
VO_{2max}, maximal oxygen consumption

A RECENT UNDERTAKING IN THE HISTORY OF MEDICINE

Nothing less than four millennia—longer if we count from the inception of Chinese medicine—went by until the fundamental mechanical principles of the heart were unveiled. Why did it take so long? The heart's mystique, perhaps, or simply the deep-rooted conservative wisdom of anyone dealing with fragile entities, such as physicians, may have prevented earlier detection. As a case in point, the influential Austrian surgeon Theodor Billroth, regarded as one of the founding fathers of modern surgery, stated in his *Handbook of General and Special Surgery* (1882): 'In my opinion, [heart surgery is] an operation approaching rather closely that point which some surgeons call prostitution of the art of surgery, others a surgical frivolity'.[1] In this line, a few years later (1896), the British surgeon and pro-vivisection campaigner Stephen Paget wrote: 'Surgery of the heart has probably reached the limits set by Nature to all surgery; no new method, and no new discovery, can overcome the natural difficulties that attend a wound of the heart'. They were wrong for almost a century. In 1801, Francisco Romero, an unbridled Spanish surgeon, performed thoracic incisions into the outermost fibrous cardiac layer (pericardium)

DOI: 10.1201/9781003486893-2

to restore normal heart function and become the first (documented) successful heart surgeon. On the other side of the Atlantic, three years before Paget's assertion (1893), the young African American surgeon Daniel Hale Williams saved the life of a moribund with a stab wound in the heart by sewing his pericardium at a very modest clinic in Chicago. These early ventures into heart surgery spur late clinical prowess and experimentation, a mandatory activity in the attempt to shed light on the convoluted products of evolution.

ERNEST STARLING, THE MASTER EXPERIMENTAL PHYSIOLOGIST

The man who first understood the fundamental inner workings of the heart was not an ordinary scientist. The British physiologist Ernest Starling (1866–1927) decisively contributed to our present understanding of how the human body—not only the heart—works. Namely, he discovered the balance of hydrostatic and osmotic forces at the capillary level (Starling's Principle), which still prevails as one of the few long-lasting lessons in medical schools around the world; he foresaw the contemporary field of endocrinology, i.e., the study of hormones; understanding of the main function of the kidney was also substantially advanced following Starling's accurate experiments and flawless observations. In between, in 1914, he made the necessary qualitative leap to grasp the essential cardiac principle, currently known as the (Starling) Law of the Heart.

One may expect a vast amount of awards and accolades over Starling's shoulders. Indeed, many of his colleagues, with far less relevant scientific contributions, were given knighthoods and Nobel Prizes. Yet, Starling received almost nothing, and he was frequently treated as an outsider. Bad luck and times of political turmoil could always be argued. Others have pointed to his character.[2, 3] To offer a glimpse, legend has it that when one of his well-esteemed colleagues, A. V. Hill, received the Nobel Prize for works in exercise physiology, the students carried Hill on their shoulders up to Starling's laboratory, at the top floor of the institute, while yelling, 'Who says he doesn't know a word of Physiology?'. 'I did, he doesn't know a damned word', Starling replied, with no sarcasm. According to his main biographer,[2] Starling was not envious; in certain situations, he just could not elude to speak his mind. He held uncompromising views, a strong sense of fairness—primarily with truth—and a passionate drive for knowledge. In his own words, physiology was 'the greatest game in the world'. Such a character inevitably called then, as now, for alliances of enemies within the flock of less capable, politically oriented and overambitious peers.

Starling experimented not with humans but with dogs, as did many of his contemporaries. They typically used the vivisection technique, which involves the dissection of anesthetized living animals. In the words of William Bayliss, Starling's closest collaborator, 'Physiology was the science of living things, and could be best taught by using living animals as subjects'. By the time that Starling and Bayliss experimented on heart function, almost 20,000 animal vivisections had already been performed in the United Kingdom. Yet, Starling and Bayliss were among the few charged with a criminal offense by the British National Anti-Vivisection Society. Despite the jury of the trial unanimously finding that the experimenters had been defamed, the anti-vivisectionists erected a bronze statue of a dog in Battersea Park, London, with the following inscription:

> In Memory of the Brown Terrier Dog done to death in the Laboratories of University College in February 1903 after having endured vivisection extending over more than two months and having been handed over from one vivisector to another till death came to his release. Also in memory of the 232 dogs vivisected at the same place during the year 1902.
>
> Men and Women of England, how long shall these things be?

One might wonder whether the current progeny of those anti-vivisectionists may hold a different opinion. A substantial fraction of them may be alive or have enjoyed extra years with their relatives thanks to effective treatments against cardiovascular disease—the number one killer in the last century. Some of those treatments would not be available without the basic knowledge acquired through experiments involving the sacrifice of non-human animals. Let us make clear that the element of cruelty is not intrinsic to the morally sound experimental enterprise.

THE LAW OF THE HEART

Eleven words suffice to convey the essential: the more the heart fills, the stronger the force of contraction. A cohesive mental display of the determinants of cardiac capacity, which is virtually to say of endurance performance, will be herein provided, including little more than the essential. The heart is comprised of two main pumps, located in the right and left halves. The right heart delivers blood to a relatively small fraction of the body: the lungs. The left heart supplies oxygenated blood to the rest of the body. Blood pumped by the left heart flows via the systemic circulation down large and small arteries, arterioles and capillaries, which have small pores where the exchange of gases and nutrients occurs. Parenthetically, the average distance between capillaries in the human body is remarkably tiny, around 0.004 cm. Despite the fact that the capillaries cannot be easily perceived by the eye, we are full of them. Capillary blood, having relatively low O_2 and high CO_2, continue flowing through the systemic veins back to the right heart, where it is pumped towards the pulmonary circulation, gathering O_2 and releasing CO_2 before returning to the left heart, thus 'closing' the double circulatory loop characteristic of mammals and birds.

Common sense instinctively projects the circulatory system as if we were moving 'forward' along with the blood flow. However, looking 'backwards' is the preferable way to understand the Law of the Heart. In fact, blood flow is not only propelled by cardiac contractions (beats). The heart is a suction pump that generates a pressure gradient driving the blood in the 'right' direction owing to the presence of venous and cardiac valves.[4] Blood in the circulation is indeed continually 'sucked' by the heart.

At this point a question may arise. How can blood be continuously flowing into the heart? Should not the returning blood flow be transitorily stopped when the heart contracts? It would do so if each half of the heart were comprised of a single chamber. Instead, both the right and left heart consist of two chambers: atrium and ventricle. Blood coming from the systemic and pulmonary circulations first enters the right and left atrium, respectively. Blood flows from the atria to the relaxed ventricles through atrioventricular valves in a predominantly passive manner driven by the pressure gradient. The filled ventricles pump the blood towards the pulmonary and systemic circulatory loops—at the location in which one loop ends (in the right or left heart), the other begins. It is the highly distensible nature of the atriums that prevents the interruption of blood flow throughout the circulatory system during ventricular contraction. Otherwise, if the heart were comprised only of ventricles, the inertia caused by intermittent blood flow would reduce the circulation and therefore cardiac pumping capacity by ~25% of its maximal values. In practical terms, without atriums, most humans would be barely capable of mild jogging.

The importance of blood flowing back to the heart, known as venous return, and the subsequent ventricular filling was progressively acknowledged by late 19th-century physiologists, Starling's predecessors. Observations from animal studies were accumulated pointing to the positive relationship of central venous pressure (a surrogate of venous return) and ventricular output, i.e., the volume of blood pumped by the ventricles per unit of time.[5] Starling, taking advantage of his experimental lucidity, e.g., via the heart–lung preparation, integrated pieces of evidence concerning cardiac and skeletal muscle to unveil the underlying principle. In his own words,

> the law of the heart is therefore the same as that of skeletal muscle, namely that the mechanical energy set free on passage from the resting to the contracted state depends on the area of 'chemically active surfaces', i.e. on the length of the muscle fibres.

Later in the article, he synthesized: 'The output of the heart is a function of its filling, the energy of its contraction depends upon the state of dilatation of the heart's cavities'.[6] Starling thus provided a mechanistic explanation, eventually confirmed several decades later with the blossoming of technological advances in molecular biology. At present, we certainly know that when the ventricles 'dilate' (expand), up to a certain length, the number of contractile protein cross-bridges within the myocardium (i.e., the cardiac muscle) is increased, thereby augmenting the force of contraction. The fundamental notion to have in mind has not changed from Starling's days: the amount of blood flowing into and expanding the ventricles is the primary determinant of their capacity to pump blood in each heart beat, known as the stroke volume, a key variable that dictates the potential for aerobic exercise capacity in humans.

HEART SIZE: WHAT THE PERICARDIUM CONCEDES

It should be no wonder, following the previous section, that heart size matters. As aforementioned, the heart of healthy sedentary adults averages 245 g in women and 331 g in men, representing less than 0.5% of body weight and volume.[7, 8] Already in the 19th century, the Swedish physician Salomon Eberhard Henschen postulated that 'big hearts win races' (1899).[9] Armed with his fingertips as the only assessment technique (chest percussion), he rightly guessed that prolonged endurance training caused enlargement of the heart, what he termed the 'athlete's heart'. Such a generalized expansion, formally known as cardiac eccentric hypertrophy, is prevalently detected in the left ventricle of endurance athletes, which regularly demand sustained high levels of blood flow to their circulatory system. Both endurance-trained women and men exhibit this feature, albeit women generally present around 10–20% smaller cardiac dimensions adjusted by body size relative to men of similar endurance training status.[10, 11, 12] In comparison with healthy sedentary individuals, the endurance athlete's heart appears to be overfilled ('inflated') by the large volume of blood that it can accommodate. Their cardiac walls are thus remarkably supple; they do not look strong at first sight, but so they are.

A large and compliant heart facilitates venous return and ventricular filling, thus leveraging the ventricular capacity to generate a large stroke volume. Cardiac pumping capacity is the product of maximal stroke volume and maximal heart rate (HR_{max}). The latter is essentially fixed for a given age, i.e., it is almost exclusively determined (reduced) by age, independently of sex and endurance training status. Accordingly, the lower ventricular filling and stroke volume in women predominantly contributes to their lower aerobic exercise capacity relative to men. Does the smaller cardiac size ultimately determine the reduced cardiac pumping capacity of women? Not necessarily. The reason is that ventricular filling in upright exercising humans is not maximal. Women and men reach their maximal functional ventricular filling and stroke volume in horizontal (supine) or slightly head-down tilt positions.[13] At any head-up tilt degree, the filling of the heart is suboptimal. That is the price to pay for bipedalism: the force of gravity limits venous return up to the heart.

In fact, *athletic* mammals with outstanding cardiac output and aerobic exercise capacity (two- to threefold higher than the aerobically fittest humans) are quadrupeds. Most of their blood volume is above the level of the heart, thus taking advantage of the effect of gravity on venous return and cardiac filling.[14] In these mammals as well as in horizontally positioned humans, the structure of the heart could be the main limiting factor. This has indeed been experimentally demonstrated in quadrupeds (dogs and pigs). Certainly, pig breeds cannot be considered as a highly fit species, yet they present with VO_{2max} levels in the range of healthy, moderately active humans. In order to test the role of heart size on cardiac and aerobic exercise capacities, Kirk Hammond and colleagues at the Division of

Cardiology of the University of California allocated female and male adult pigs into two main experimental groups.[15] The first one had their chest open (thoracotomy) and their pericardium cut, thereby allowing the heart to be unrestrained from the external fibrous protective layer. The second group also underwent thoracotomy, but their pericardium was left intact, thus serving as a control for the potential confounding effect of thoracotomy. Both groups performed an incremental exercise test before the experimental procedure and 2–3 weeks after to allow for recovery from surgery.

The results were astonishing. Pericardiectomy induced a 33% increase in left ventricular volume. Hence, the filling of the heart was augmented by one third after removing the constraint naturally enforced by the pericardium. Likewise, striking proportional increases in maximal stroke volume (35%), cardiac output (29%) and VO_{2max} (31%) were observed after pericardiectomy. In contrast, no improvements were detected in the control thoracotomy group with intact pericardium: thoracotomy per se reduced stroke volume (–12%), cardiac output (–12%) and VO_{2max} (–21%). Indeed, when taking into account the negative effects of thoracotomy, the independent effects of pericardiectomy are still more overwhelming, resulting in ~50% improvements in stroke volume and VO_{2max}. Therefore, pericardiectomy per se *upgraded* sedentary pigs having a typical VO_{2max} observed in healthy, moderately active humans (48 mL/min/kg) to that of elite endurance runners (~72 mL/min/kg). Similar results have been reported in dogs.[16] Furthermore, there is also anectodical evidence of pericardiectomy in greyhound racing,* where the stakes mainly rely on the dog's cardiac pumping capacity. The gambler seems to appreciate (better that some academics) the major limitation enforced by the pericardium on cardiac filling, output and thereby O_2 delivery and VO_{2max} in mammals characterized by a body position (quadrupedal) in that venous return is not impeded by gravity.

The role of the pericardium in aerobic exercise performance in humans remains unexplored. Conforming to basic ethical guidelines, straightforward experimental evidence from pericardiectomy studies is not available in healthy humans. The pericardium may be there for a good evolutive reason—judicious people may argue. The thin, albeit strong, double-walled outer layer that surrounds the heart seemingly protects and attenuates friction by enclosing the

* Personal communication.

myocardium (the middle contracting layer) in a serous ('lubricating') fluid. Notwithstanding, can humans survive without a pericardium? Apparently, yes. The congenital absence of the pericardium does not seem to alter human life expectancy.[17] Moreover, in patients with stiffened cardiac tissues due to inflammatory processes, the removal of the pericardium is a legitimate clinical option as long as adjacent organs (lungs, diaphragm) are not affected. Likewise, a minimal opening of the pericardium is beneficial in patients undergoing cardiac surgery in order to vastly reduce (67% decrement) the pressure required to fill the left ventricle.[18] Furthermore, pericardiectomy has been recently recommended in leading cardiology journals to treat patients presenting with high filling pressures leading to stiff and concentrically hypertrophied hearts[19]—the opposite cardiac phenotype of endurance athletes. In fact, concentric hypertrophy entails cardiac internal growth (thickened walls) without overall enlargement, thereby curtailing cardiac filling and pumping capacity.

The pericardium's narrative may gain further relevance in women (Figure 2.1). Even in the absence of morbid conditions, the female heart is prone to develop structural modifications resulting in cardiac concentric hypertrophy.[20] Women represent two thirds of the aforementioned patients who might benefit from pericardiectomy.[21,22] Healthy women undergoing endurance training also present distinct cardiac adaptations compared with similarly trained men, entailing limited improvements in cardiac filling and output.[23] Women's specific adaptability to endurance training stimuli along with potential 'natural' interventions to maximize cardiac eccentric expansion will be developed in Chapter 6 in conjunction with other relevant sex-specific responses to exercise training. Yet, the hypothetical role of pericardial constraint on women's prevalent cardiac remodeling is uncertain at present.

Recent evidence from our laboratory provides a mixed picture of the potential role of the pericardium on the limitation of cardiac capacity in women.[24] Using a placebo-controlled and crossover study design, blood volume was incremented by 10% via the intravenous infusion of a plasma volume expander in healthy young women and men matched by age and moderate physical activity (in total and specific to endurance exercise). At rest, the expansion of blood volume proportionally increased (+10%) cardiac filling in men, but not in women. During incremental cycling exercise from moderate to maximal exercise intensities, cardiac filling was again increased (+10% on average) in men, whereas this increment

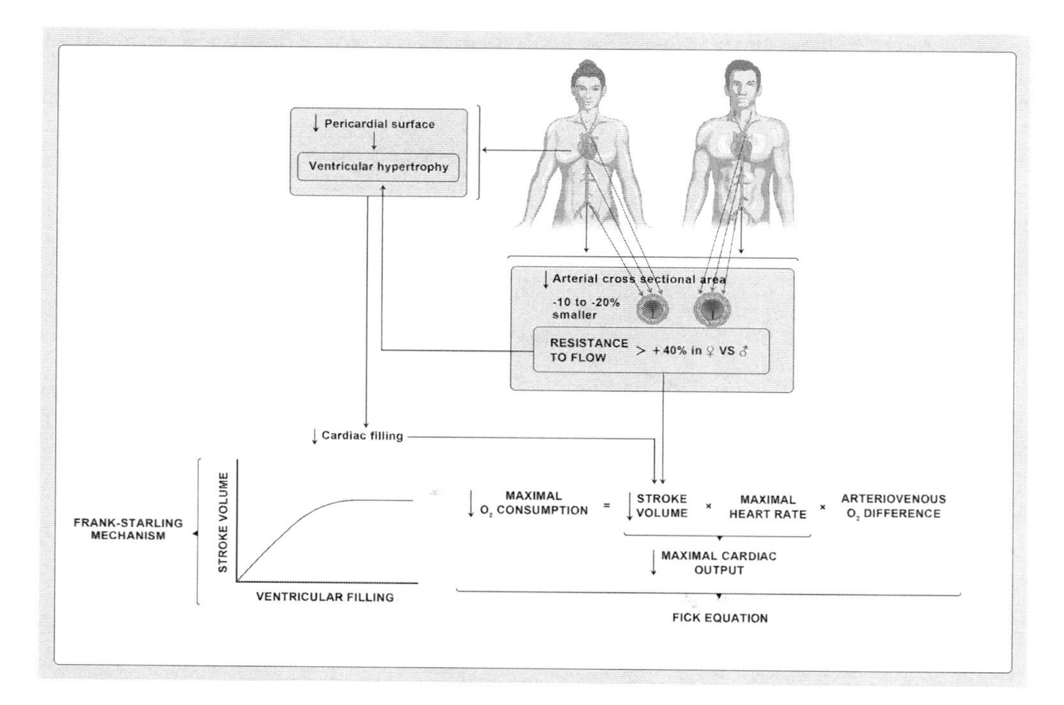

Figure 2.1 Main sex differences in the circulatory system affecting cardiac and aerobic capacities. The man and woman in the figure have the same body weight.

(+10%) was observed only at high intensities (≥80% HR_{max}) in women. Less pericardium elasticity, due to its possibly being closer to the ultimate tensile strength in women, might, at least in part, explain their delayed cardiac filling. At peak effort women were able to fill their heart in proportion to the increment in blood volume, suggesting no limitation at very high cardiac filling pressures, which nonetheless could hardly be maintained for more than a few minutes.

A NOTE ON THE MIND OF ELITE PERFORMERS

Extrapolation from the clinic to the athletic arena, while often being groundless, is expected among individuals in pursuit of performance enhancement. The potential outstanding effects of pericardiectomy may surely appear tentative to win-at-all-cost characters. Would any healthy woman or man be willing to have its own pericardium cut in the attempt to win sporting events, notwithstanding the lack of specific supporting evidence regarding the acute effects and long-term consequences? As outrageous at it may seem, the intricated history of doping and elite performance has clearly exposed the bloody and ruthless psyches of certain athletes, coaches and medical doctors. What they could be able to do to secure triumph can hardly be underestimated.

WOMEN'S HEART: MORE TO IT THAN CARDIAC SIZE AND PERICARDIAL CONSTRAINT

Regardless of advances in healthcare technology, the simple measurement of arterial blood pressure prevails as an indispensable cardiovascular and overall health marker. Blood pressure in large (non-pulmonary) arteries reflects the force generated by the heart to pump blood through the systemic circulation. Typical values in resting normotensive European American individuals are ~120 millimeters of mercury (mm Hg) for the highest (systolic) and ~80 mm Hg for the lowest (diastolic) pressure during a cardiac cycle. By convention, these numbers are relative to the given external atmospheric pressure. The unit, mm of Hg, refers to the force exerted by the height of a column of mercury, which is a stable and very dense fluid relative to water (1 mm of Hg is equivalent to 13.6 mm of water). Considering that blood and water densities are nearly equal, a puncture in a large artery of a normotensive individual may result in a blood spurt 163 cm tall, approximately the average height of women. Such a high pressure must be generated by the heart to deliver blood to the body tissues at rest. During exercise, the values are increased two- to threefold. Hence, it is better not to puncture an artery at maximal exercise. It can be hinted that even slight chronic sex differences in blood pressure may lead to substantial sex-specific cardiac adaptations, as the heart must continually overcome, beat after beat, the resistance to circulate blood imposed by the vasculature.

For the majority of the adult lifespan, healthy European American women and men present with similar arterial blood pressures at rest.[25] Exercise blood pressure is also not different between sexes.[26] The sex parity in arterial blood pressure entails inescapable consequences, which will be readily grasped after a few basic hemodynamic relationships are defined. Namely, systemic vascular resistance is the force of friction that opposes blood flow, intuitively and mainly determined by the overall opening and length of blood vessels. Mathematically, systemic vascular resistance is directly proportional to mean arterial blood pressure and inversely to systemic blood flow (cardiac output): the ratio of mean arterial blood pressure and cardiac output defines systemic vascular resistance. Such a ratio must diverge between sexes, as women and men have different cardiac outputs for a given mean arterial blood pressure.

Cardiac output at rest approximately amounts to 5 liters per minute in adult women from 20 to 80 years of age and normal body weight, half a liter less than age-matched men with ~20% higher body weight.[27] The sex gap in cardiac output is magnified during exercise, plausibly underlied by enhanced structural (pericardial) and/or functional limitations to ventricular filling in women.[26] Sedentary and moderately active men deliver approximately 4 to 5 liters of blood more per minute at maximal exercise than women matched by age and physical activity levels, despite having similar arterial blood pressures.[26] At the extreme of fitness, a large gap in maximal cardiac output has been reported between female and male Olympic medalists in endurance events (~35 vs. ~45 L/min).[28] Normalized by body weight, female elite endurance athletes still exhibit around a 15% lower maximal cardiac output compared with male counterparts.[28] Across the fitness spectrum, the female heart pumps less blood while generating the same force (blood pressure) as male counterparts. Consequently, the resistance to deliver blood to the systemic circulation must be greater in women.[26, 29, 30] This may additionally contribute to prevalent cardiac concentric hypertrophy in women.[31] As far as the analogy holds, women tend to develop a bulky yet small heart, akin to the body of small-framed individuals devoted to resistance training.

Why is there a higher resistance to circulate blood in women? The answer, at least its physical basis, has been known for almost two centuries. In 1846, the French physiologist Jean-Louis-Marie Poiseuille outlined the principles of laminar flow based on the properties of non-turbulent flow through regular pipes, which fairly resembles the circulation in blood vessels.[32] Poiseuille's experiments in animals and small capillaries revealed the main determinant of the friction opposing blood flow: the cross-sectional internal area of blood vessels. Notably, blood flow resistance is proportional to the fourth power of the diameter of a vessel. The exponential feature of this physical relationship is crucial. A two-fold difference in vessel diameter entails a 16-fold difference in blood flow resistance.

The diameter of the primary artery delivering blood to the systemic circulation, i.e., the aorta, is consistently smaller (–10%) in women relative to men throughout the adult lifespan.[33] The sex difference in aortic diameter is even larger (>10%) in elite endurance and/or strength-trained athletes.[34] Similarly, the diameter of downstream muscular arteries in the legs and arms is reduced (–10% to –20%) in women compared with men (Figure 2.1).[35, 36] Likewise, arteries perfusing the heart (coronary arteries) possess a smaller (–10% to –15%) in women than in men, even when adjusted by body size and cardiac mass.[37] While arterial size is not a static characteristic and could be in theory compensated by vasodilation, the evidence indicates that the overall systemic vascular resistance during dynamic exercise, from moderate to high intensity levels, is largely augmented (> +40%) in women relative to men, which approximately corresponds to Poiseuille's predicted impact of sex differences in blood vessel diameter.[26]

Collectively considered, compared with men, the smaller heart of women can generate a similar amount of force, i.e., blood pressure; yet the reduced female blood vessel diameter exponentially increases the resistance to flow, resulting in markedly reduced cardiac pumping capacity. Women would need to produce far higher blood pressure (>500 mm Hg) to close the cardiac output gap with men. Such a blood pressure is not even reached by heavy-weight lifters during maximal efforts, which entails, frightfully, the risk of surpassing the boundary of arterial breaking strength.[38] In addition, the coronary artery capacity to deliver blood and thereby O_2 to cardiac muscle fibers seems to be structurally curtailed in women. So must be the theoretical maximal contractile capacity of the heart, which almost exclusively depends on aerobic energy production. Regardless, the frequently similar blood pressure developed by the smaller heart of women entails a greater stress per muscle fiber. This may stimulate, along with pericardiac constraint, concentric hypertrophy, in turn restricting cardiac filling and output, closing the maladaptive loop induced by reduced blood vessel diameter.

In fact, women do substantially better in terms of cardiovascular health when their heart is exposed to blood pressure levels lower than those recommended by general guidelines.[39] The systolic blood pressure at which women start showing a significant risk of cardiovascular disease, the current leading of cause of death worldwide, is 100–109 mm Hg, while that of men is 130–139 mm Hg.[39] Hence, most women may chronically overload their heart, even though they generally live longer than men, plausibly as a result of compounding biological, behavioral and environmental factors—a longer life with an inherently limited capacity to deliver O_2 and therefore exercise capacity due to the laws of physics and the parcourse of human evolution. Yet, the cardiovascular system cannot be understood in isolation (heart and vessels). We know that circulatory physiology inherently depends on blood, a fascinating fluid in which we must be immersed to complete this story.

REFERENCES

1. Billroth T. *Handbuch der allgemeinen und speciellen Chirurgie:mit einschluss der topographischen anatomie, operations-und verbandlehre*. Ferdinand Enke Verlag; 1882:63–164.
2. Henderson J. *A life of Ernest Starling*. Academic Press; 2005.
3. Henriksen JH. Ernest Henry Starling (1866–1927): The scientist and the man. *J Med Biogr.* 2005;13(1):22–30.
4. Anderson RM, Fritz JM, O'Hare JE. The mechanical nature of the heart as a pump. *Am Heart J.* 1967;73(1):92–105.
5. Katz AM. Ernest Henry Starling, his predecessors, and the 'Law of the Heart'. *Circulation.* 2002;106(23):2986–2992.
6. Patterson SW, Piper H, Starling EH. The regulation of the heart beat. *J Physiol.* 1914;48(6):465–513.

7. Molina DK, DiMaio VJ. Normal organ weights in men: Part I—the heart. *Am J Forensic Med Pathol.* 2012;33(4):362–367.
8. Molina DK, DiMaio VJ. Normal organ weights in women: Part I—the heart. *Am J Forensic Med Pathol.* 2015;36(3):176–181.
9. Rost R. The athlete's heart. Historical perspectives—solved and unsolved problems. *Cardiol Clin.* 1997;15(3):493–512.
10. Pfaffenberger S, Bartko P, Graf A, et al. Size matters! Impact of age, sex, height, and weight on the normal heart size. *Circ Cardiovasc Imaging.* 2013;6(6):1073–1079.
11. Rowland T, Roti M. Influence of sex on the 'Athlete's Heart' in trained cyclists. *J Sci Med Sport.* 2010;13(5):475–478.
12. Patel HN, Miyoshi T, Addetia K, et al. Normal values of cardiac output and stroke volume according to measurement technique, age, sex, and ethnicity: Results of the world alliance of societies of echocardiography study. *J Am Soc Echocardiogr.* 2021.
13. Bundgaard-Nielsen M, Sørensen H, Dalsgaard M, Rasmussen P, Secher NH. Relationship between stroke volume, cardiac output and filling of the heart during tilt. *Acta Anaesthesiol. Scand.* 2009;53(10):1324–1328.
14. Uhrikova I, Lacnakova A, Tandlerova K, et al. Haematological and biochemical variations among eight sighthound breeds. *Aust Vet J.* 2013;91(11):452–459.
15. Hammond HK, White FC, Bhargava V, Shabetai R. Heart size and maximal cardiac output are limited by the pericardium. *Am J Physiol.* 1992;263(6 Pt 2):H1675–H1681.
16. Stray-Gundersen J, Musch TI, Haidet GC, Swain DP, Ordway GA, Mitchell JH. The effect of pericardiectomy on maximal oxygen consumption and maximal cardiac output in untrained dogs. *Circ Res.* 1986;58(4):523–530.
17. Shah AB, Kronzon I. Congenital defects of the pericardium: A review. *Eur Heart J Cardiovasc Imaging.* 2015;16(8):821–827.
18. Borlaug BA, Schaff HV, Pochettino A, et al. Pericardiotomy enhances left ventricular diastolic reserve with volume loading in humans. *Circulation.* 2018;138(20):2295–2297.
19. Borlaug BA, Reddy YNV. The role of the pericardium in heart failure: Implications for pathophysiology and treatment. *JACC Heart Fail.* 2019;7(7):574–585.
20. Regitz-Zagrosek V, Kararigas G. Mechanistic pathways of sex differences in cardiovascular disease. *Physiol Rev.* 2017;97(1):1–37.
21. Diaz-Canestro C, Montero D. Female sex-specific curtailment of left ventricular volume and mass in HFpEF patients with high end-diastolic filling pressure. *J Hum Hypertens.* 2021;35(3):296–299.
22. Scantlebury DC, Borlaug BA. Why are women more likely than men to develop heart failure with preserved ejection fraction? *Curr Opin Cardiol.* 2011;26(6):562–568.

23. Diaz-Canestro C, Montero D. The impact of sex on left ventricular cardiac adaptations to endurance training: A systematic review and meta-analysis. *Sports Med.* 2020;50(8):1501–1513.
24. Guo M, Diaz-Canestro C, Montero D. The Frank-Starling mechanism is not enough: Blood volume expansion prominently decreases pulmonary O_2 uptake. *Mil Med Res.* 2024;11(1):43.
25. Wills AK, Lawlor DA, Matthews FE, et al. Life course trajectories of systolic blood pressure using longitudinal data from eight UK cohorts. *PLoS Med.* 2011;8(6):e1000440.
26. Diaz-Canestro C, Pentz B, Sehgal A, Montero D. Sex differences in cardiorespiratory fitness are explained by blood volume and oxygen carrying capacity. *Cardiovasc Res.* 2021;118(1):334–343.
27. Rusinaru D, Bohbot Y, Djelaili F, et al. Normative reference values of cardiac output by pulsed-wave Doppler echocardiography in adults. *Am J Cardiol.* 2021;140:128–133.
28. Lundby C, Robach P. Performance enhancement: What are the physiological limits? *Physiology (Bethesda).* 2015;30(4):282–292.
29. Diaz-Canestro C, Pentz B, Sehgal A, Montero D. Sex differences in cardiorespiratory fitness are explained by blood volume and oxygen carrying capacity. *Cardiovasc Res.* 2022;118(1):334–343.
30. Diaz-Canestro C, Pentz B, Sehgal A, Yang R, Xu A, Montero D. Lean body mass and the cardiovascular system constitute a female-specific relationship. *Sci Transl Med.* 2022;14(667):eabo2641.
31. Diaz-Canestro C, Ng HF, Yiu KH, Montero D. Reduced lean body mass: A potential modifiable contributor to the pathophysiology of heart failure. *Eur Heart J.* 2023;44(16):1386–1388.
32. Poiseuille J. Recherches expérimentales sur le mouvement des liquides dans les tubes de très petits diamètres. In: *Mémoires présentés par Divers Savants à l'Académie Royale des sciences de l'Institut de France.* Imprimerie Royal, Paris, 1846:433–544.
33. Mao SS, Ahmadi N, Shah B, et al. Normal thoracic aorta diameter on cardiac computed tomography in healthy asymptomatic adults: Impact of age and gender. *Acad Radiol.* 2008;15(7):827–834.
34. Boraita A, Heras ME, Morales F, et al. Reference values of aortic root in male and female white elite athletes according to sport. *Circ Cardiovasc Imaging.* 2016;9(10):e005292.
35. Sandgren T, Sonesson B, Ahlgren R, Lanne T. The diameter of the common femoral artery in healthy human: Influence of sex, age, and body size. *J Vasc Surg.* 1999;29(3):503–510.
36. van der Heijden-Spek JJ, Staessen JA, Fagard RH, Hoeks AP, Boudier HA, van Bortel LM. Effect of age on brachial artery wall properties differs from the aorta and is gender dependent: A population study. *Hypertension.* 2000;35(2):637–642.
37. Hiteshi AK, Li D, Gao Y, et al. Gender differences in coronary artery diameter are not related to body habitus or left ventricular mass. *Clin Cardiol.* 2014;37(10):605–609.

38. MacDougall JD, Tuxen D, Sale DG, Moroz JR, Sutton JR. Arterial blood pressure response to heavy resistance exercise. *J Appl Physiol.* 1985;58(3):785–790.
39. Ji H, Niiranen TJ, Rader F, et al. Sex differences in blood pressure associations with cardiovascular outcomes. *Circulation.* 2021;143(7):761–763.

Chapter 3

Blood

Abbreviations:

HR_{max}, maximal heart rate
O_2, oxygen
PFCP, primary familial and congenital polycythemia
VO_{2max}, maximal oxygen consumption

THE IMPORTANCE OF BLOOD VOLUME FOR EXERCISE CAPACITY

Healthy men of normal body weight and height generally have little more than 5 liters of blood.[1–3] Within the same total body volume, highly trained endurance athletes may squeeze 8 to 9 liters of blood.[4] The parallelism between endurance training and augmented blood volume was first explicitly appreciated by Swedish researchers in the aftermath of World War II.[5] Thereafter, greater blood volume has been consistently found in cross-sectional studies including endurance athletes, presenting with up to 40% increments versus untrained peers.[4, 6–8] Similar fitness-related differences are observed in women, who, for a given endurance training status, commonly present 5–10% less blood per kg of body weight than men (Figure 3.1).[1, 4] Overall, both women and men with outstanding cardiac pumping capacity invariably present with elevated blood volume. Hence, blood volume is strongly correlated with cardiac and aerobic capacities.[4] Yet, correlation does not mean causation. To prove cause and effect, experimental studies manipulating the variable of interest, i.e., blood volume, were indispensable.

DOI: 10.1201/9781003486893-3

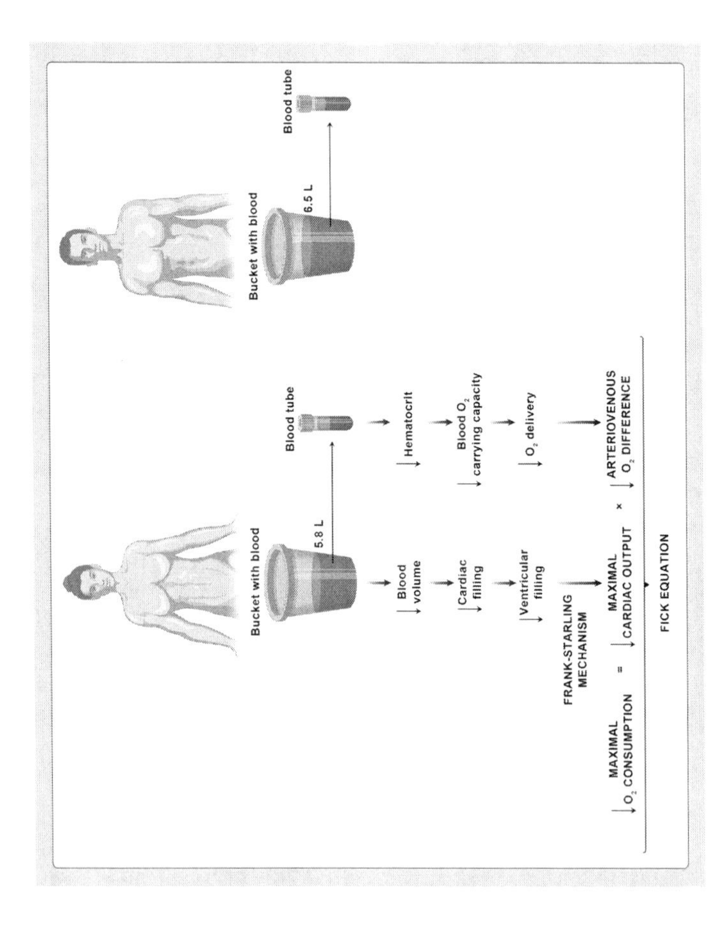

Figure 3.1 Sex differences in blood affecting cardiac and aerobic capacities. The man and woman in the figure have the same body weight.

Among the potential determinants of cardiac function, only a few can be safely manipulated in humans. Indeed, no human ethical committee would approve of altering the intrinsic structure of the heart for experimental purposes. In contrast, blood is nowadays a readily manipulable variable in healthy individuals. Its role in cardiac and aerobic capacities can therefore be neatly addressed, for instance via phlebotomy. Herein, the debilitating effects of blood withdrawal on endurance performance were already noticed in the 1940s, when Springfield College ruled that 'no man may place his name on the blood donors list while actively engaged in a varsity sport'.[9] To date, multiple studies have confirmed the substantial reduction in aerobic capacity induced by a standard blood donation, equivalent to ~450–500 mL of blood.[10–30] Two to three weeks are generally required for the recovery of aerobic capacity, which coincides with the necessary time span for the approximate restitution of the total volume of circulating red blood cells, which along with plasma volume constitute nearly all the blood volume.[24, 29, 31]

Historically, most studies in exercise physiology have included young men. The topic of this chapter is no exception. Recently, the effects of a standard blood donation on cardiovascular function and aerobic capacity were comprehensively investigated in healthy women throughout the adult lifespan.[32] Similar to men, blood withdrawal proportionally impaired cardiac filling and stroke volume in women. Consequently, systemic O_2 delivery at any percentage of VO_{2max} is reduced in women according to the magnitude of hypovolemia. Given that O_2 delivery is decreased at any given heart rate following blood withdrawal, women and men undergoing prolonged physical effort at a given heart rate must perform at a lower absolute exercise intensity after phlebotomy, a pertinent observation for physical activity recommendations based on heart rate monitoring.[33] Also relevant are the implications for women undergoing surgery, since the same procedures performed in either sex commonly result in a comparable loss of blood volume.[34] Likewise, the recovery of aerobic capacity may be delayed in women for a certain absolute amount of blood loss, which constitutes a higher fraction of blood volume in women than men matched by physical activity.[4] In hindsight, the early directive of Springfield College should have been not exclusive for men, yet more appropriate for women.

To the surprise of many, when blood volume per kg of body weight is precisely matched between healthy young women and men via blood withdrawal in the latter, sex differences in cardiac output during exercise are no longer present.[35] If, in addition to blood volume normalization, blood O_2 carrying capacity is also matched (reduced) in men to the level of women, a 19% sex gap in VO_{2max}, which is by no means negligible, is completely abrogated.[35] The striking contribution of blood attributes to the so-called sex dimorphism in aerobic capacity might not be generally expected among exercise physiologists. Certainly, blood is not the only variable along the O_2 transport and utilization chain known to possess sex-specific characteristics.[36–38] Experimental evidence is frequently stubborn relative to our expectations: not every sex difference along the O_2 cascade, from the lungs to the mitochondria in skeletal muscle, explains the gap in VO_{2max} between sexes.

The overwhelming impact of blood on aerobic exercise capacity in humans portrays powerful information. As previously stated, 'the ultimate goal of any area of Physiology is to discover the fundamentals of how a given function works, thus empowering to modify outcomes as desired'.[39] Power to effectively modify outcomes requires understanding; otherwise, little can be realized, which is appreciated by anyone (good-hearted or not) resolutely willing to succeed, as it will be manifest in this chapter.[40]

Coming back to physiology, blood is intimately related to the law of the heart—the more the heart fills, the stronger the force of contraction.[41] The amount of blood flowing into and thereby expanding the ventricles, i.e., the preload, is well beneath the upper (pericardial) limit in the upright position, such that any change in ventricular filling alters stroke volume proportionally.[42] Cardiac filling depends on a pressure gradient, from large veins into the right atrium. This gradient, in the absence of major structural cardiovascular abnormalities, is mainly dictated by venous compliance and the filling of the circulatory system, reflected by blood volume.[43] As indicated in Chapter 2, optimal cardiac filling in humans occurs in horizontal or slightly head-down tilt positions.[42] Therefore, blood volume determines, not exclusively but predominantly, stroke volume at rest as well as during exercise in upright and semi-upright body positions.[1, 32, 35, 44] The greater the blood volume, for a given body size, the larger the stroke volume and thereby cardiac pumping capacity in non-supine or head-down body positions, which comprise most of the spectrum of physical activities involving aerobic exercise. In supine or head-down body positions, blood volume may still play a role: the greater the filling of the circulatory system, the larger the average internal cross-sectional area of blood vessels, and thereby the lower the resistance to circulate blood.

The importance of blood volume for the function of the heart during exercise can hardly be overstated. This is obvious when potential contributions of alternative independent factors are experimentally assessed. In principle, HR_{max} is a primary determinant of cardiac output that is not affected by blood volume. The Danish research group led by Stefan Mortensen tested the counter-intuitive hypothesis that heart rate during exercise does not affect cardiac output. To this end, they inserted catheters in the heart that augmented heart rate by 20 beats per minute over normal levels from rest until maximal upright exercise (cycling and one-legged knee extension) in healthy young men.[45] Cardiac output was unaltered throughout exercise despite the supraphysiological increase in heart rate. Conforming to the fact that cardiac output is the product of heart rate and stroke volume, the latter was reduced in proportion to the experimental increase in heart rate. Plausibly, increases in heart rate beyond physiological levels limit the time required to fill the ventricles between heart beats. This study demonstrated that heart rate does not determine cardiac output during exercise in young men. Whether such a landmark finding can be extrapolated to women or older individuals (who must present lower HR_{max} due to older age), is still an open question. Yet, stroke volume certainly determines cardiac output and exercise capacity in any human being, hence the underlying importance of a major determinant of cardiac filling such as blood volume and its fluid-nature, amenable to *modification.*

HOW MUCH IS TOO MUCH BLOOD FOR THE HEART OF ELITE ATHLETES?

Certainly, a limit exists in blood volume, which becomes manifest in pathological conditions. For instance, in the presence of heart failure, multiple endocrine axes are upregulated, which increases fluid retention and plasma volume as well as erythropoiesis and the number of circulating red blood cells.[46] In this manner, the body attempts to compensate for the intrinsic failure of the heart via augmented blood volume and thereby cardiac filling, i.e., exploiting the law of the heart. The issue, however, is that such a response frequently overcompensates beyond the optimal filling of the failing heart. Astronomical levels of extracellular fluid—more than 30 liters, or 450 mL per kg of body weight—have been reported in heart failure patients, nearly tripling the normal values in the healthy population.[47] Blood volume may exceed 100 mL per kg in these patients, a ~1.5-fold increase compared with healthy moderately active individuals.[48] The massive fluid overload results in generalized edema and very high cardiac wall stress, leading to impaired cardiac function, ischemia and dyspnea.[49] Hence, the fluid overload response to heart failure may yield to antithetical outcomes relative to the *intended* ones.

In logical contrast, the healthy heart of an endurance athlete withstands very high levels of blood volume, with maximal values in Olympic champions slightly over ~100 mL of blood per kg.[4] Moreover, anecdotal evidence, albeit plenty of it, indicates that still higher, supraphysiological blood volume levels can be accommodated in the circulatory system to the benefit of endurance performance.[50] In his eye-opening book *Blood Sports: The Inside Dope on Drugs in Sport*, the anti-doping expert Robin Parisotto stated: 'Transfusing blood in order to improve sporting performance only gained prominence in the 1970s. The outcomes of innumerable races were decided by one simple question: who'd transfused the most blood?'[50] In a comparison with other doping practices and athletes, Parisotto wrote:

> while 'power' athletes popped pill after pill throughout the 1970s and '80s, anabolic steroids offered only minor benefits to their endurance counterparts—cyclists, long-distance runners and swimmers. What these elite sportsmen and women needed was stamina, not strength. What they needed was more blood.[50]

Far more blood. Approximately one liter of blood is transfused before any major competition according to confiscated doping diaries of recently active elite cyclists.[40] If the competition lasts several weeks, as for instance the Tour de France, several additional units (~450 mL) of blood bags are transfused during the race.[40] The impact of a two-unit blood transfusion is massive, about a 20% increase in endurance (cycling, running) performance measured in controlled conditions.[12, 51] In practical terms, defeat is extremely unlikely for a blood-doped endurance athlete competing against 'clean' ones.

The crude reality is that not many perfectly 'clean' athletes must remain at the elite level, based on the high prevalence of abnormal blood profiles.[52] Such a situation does not seem to be reversed, contrary to recurring mass media claims.[53, 54] Indeed, main blood doping methods are still not directly detected by established anti-doping methods.[40, 55] It is only natural that skepticism grows within and around elite sports. In a field in which speaking out is often punished, the assertions of the Danish physiologist Carsten Lundby are exemplary: 'Citius, altius, fortius—the Olympic motto—is supposed to be reached solely by natural talent, proper training and diet. How true is that today? Probably not true at all'.[40] He follows with greater skepticism, but directed towards the system: 'How can it be so bad with all the anti-doping activities and controls taking place? Is it really worth all the efforts and expenses?'. Eventually, the general public—i.e., the client who ultimately finances elite sports (directly and indirectly via taxes)—should confront the dilemma.

PREGNANCY DOPING: WHEN THE OUTRAGEOUS MIGHT BECOME PREVALENT

In 1988, Risto Erkkola, a Finnish physician specializing in obstetrics and gynecology, allegedly stated: 'now that drug testing has become routine, pregnancy has become the favourite way of getting an edge on competitors'.[56] A decade later, Paul-Erik Paulev, a Danish professor of physiology at the University of Copenhagen, wrote the following in his *Textbook in Physiology and Pathophysiology*: 'in some countries female athletes have become pregnant for 2–3 months, in order to improve their performance just following an abortion'.[57] In between (1994), one of the largest media broadcasters in Germany, the RTL Group, featured a documentary in which Olga Kovalenko, a multiple gold medal winner from the Soviet gymnastics team, confirmed the practice of 'pregnancy doping'.[58] The sensationalist story included coaches forcing the gymnasts to get pregnant (or directly inseminating them) a few weeks before the competition, which was preceded by an induced abortion. The story still had another extravagant chapter. In 2004, Kovalenko claimed that she did not made those statements, an imposter apparently did—there is no available footage of the documentary. Yet, she brought only a Russian magazine to court. No legal action was taken against RTL or other worldwide media that echoed the news (e.g., *Sports Illustrated*, *Daily Telegraph*, *The Times*). Strange, to say the least. More recently, in 2017, the US track and field star Sanya Richards-Ross described in the book *Chasing Grace: What the Quarter Mile Has Taught Me about God and Life* that she had an induced abortion 1 day before leaving for the 2008 Beijing Olympics, where she won gold and bronze medals. She went further in an interview with *Sports Illustrated Now*:

> The truth is it's an issue [abortion before competition] that is not really talked about, especially in sports, and a lot of young women have experienced this. I literally don't know any female track and field athlete who hasn't had an abortion.[59]

Beyond looking into proof of pregnancy doping, we should ponder whether it makes sense physiologically. Blood volume increases dramatically with pregnancy. About an extra 1500 mL of blood are gained at the end of pregnancy, comprising both plasma and red blood cells.[60] In parallel, the size of the heart is proportionally increased along with the expansion of blood volume.[61] In the weeks following childbirth, however, the woman's body retains a substantial increase in body weight (>5 kg) compared with normal weight (pre-pregnancy). This combined with the inability to train hard during mid and late pregnancy precludes any immediate athletic benefit after delivery. If the increase in blood volume and subsequently enhanced cardiac filling and output remains for several weeks or months, then the advantage may be manifest provided that normal body weight and general fitness are re-established. Nonetheless, the majority of the aforementioned Anecdotal accounts 'recommend' stopping pregnancy (abort) in the first trimester, around the 8th to the 12th week. At this early stage, a moderate increase in blood volume, equivalent to one unit of blood, along with augmented heart volume are present;[60, 61] meanwhile, training stimuli do not need to be affected, particularly if the final outcome is abortion. Body weight gain can also be largely attenuated during this initial (three months) period. Taken together, the 'early' type of pregnancy doping might have a valid physiological basis; no need to elucidate its moral basis.

MORE BLOOD, BUT NOT IN ISOLATION

Endurance training induces blood volume expansion, which leads to increased cardiac pumping capacity, VO_{2max} and endurance performance. The primary role of blood volume in this sequence is theoretically supported by the Frank-Starling mechanism and the empirical abolishment of maximal cardiac output, VO_{2max} and performance improvements when the training-induced gain in blood volume is negated.[44, 62, 63] Yet, the role of blood volume has been mainly established via the 'negative' (blood withdrawal) sequence. Would blood volume, in isolation, explain the 'positive' sequence? Specifically, what if some non-hematological adaptations were concurrently needed for the blood volume expansion-induced increment in maximal cardiac output to translate into enhanced VO_{2max} and endurance performance? Recent experimental evidence from our laboratory (mentioned in Chapter 2) suggests so. Acute blood volume expansion (10% increment, similar to that elicited by endurance training programs)[64] via the infusion of a plasma volume expander increased cardiac filling and output during incremental cycling exercise in healthy, moderately trained women and men; whereas, unexpectedly, VO_{2max} and the maximal workload were largely reduced (–10%) irrespective of sex. These results were explained by a concomitant peripheral vasodilation and marked decrement of O_2 extraction during exercise with blood volume expansion. Hence, the gain in maximal cardiac output (blood flow) was not directed towards the active (high O_2-consuming) muscle fibers. To this end, arterial blood pressure during exercise must be increased, a common adaptation to endurance training,[65, 66] which is absent when blood volume is augmented in isolation. Therefore, complementary adaptations modifying the systemic regulation of arterial blood pressure must coexist with blood volume expansion in order that the enhanced cardiac pumping capacity and arterial (O_2-rich) blood can be directed towards the vessels irrigating the working muscles. Nonetheless, until further evidence is available, it cannot be discarded that prolonged expansion of blood volume per se (in isolation) eventually stimulates adaptations in baroregulation required to optimize blood flow distribution. These findings also imply that the well-established enhancement of VO_{2max} and endurance performance following blood doping (i.e., augmenting the number of circulating red blood cells, not just plasma) may be mostly explained (94%)[12] by the concomitant increase in blood O_2 carrying capacity, a fundamental variable upon we will focus next.

THE NATURAL BLOOD-DOPED ATHLETE AND THE OPTIMAL O_2 CARRYING CAPACITY

Thus far, we have underlined the hemodynamic prominence of blood. The content, in addition to the volume that contains it, is also crucial. Nothing is as essential in the blood of complex living beings as a metal-containing protein with high affinity for O_2. In humans, and almost all vertebrates, this precious protein, known as hemoglobin, is sheltered inside red blood cells. Hemoglobin comprises four iron atoms with high affinity for O_2. Hence, each molecule of hemoglobin binds to four molecules of O_2. With the knowledge of the molecular weight of hemoglobin and its normal hemoglobin concentration in blood in humans (13–15 gr per dL of blood), we can determine the volume of O_2 that can be transported in the circulation: around 200 mL of O_2 per liter of blood. When the oxygenated blood perfuses the tissues in need of O_2, hemoglobin loses part of its O_2 affinity due to tissue-related physical and biochemical alterations, facilitating the release of O_2. The opposite (increased affinity) occurs in the lungs, where partially deoxygenated blood is exposed to relatively high O_2 concentrations in the alveoli—as befitting as physiology can be. Without hemoglobin, every liter of blood could hold only ~1.5 mL of O_2, dissolved in the plasma. Otherwise stated, hemoglobin enables substantial O_2 delivery and aerobic metabolism, i.e., life, in vertebrates. Herein the importance of hemoglobin concentration, which determines blood O_2 carrying capacity and thereby the hematological potential to deliver O_2 to the body.

Surprising as it may seem, endurance athletes are not characterized by higher hemoglobin concentration than healthy sedentary individuals.[40] There is an exception to this general statement: athletes presenting with a rare genetic variant leading to hypersensitivity of the erythropoietin receptor, known as primary familial and congenital polycythemia (PFCP).[67] This *natural* condition, PFCP, is associated with up to 50% increments in blood O_2 carrying capacity despite circulating erythropoietin (the main hormone that regulates the formation of red blood cells and thereby Hb concentration) is not increased above normal levels.[67] In this respect, the success of Eero Mäntyranta, a multiple gold medal winner diagnosed with PFCP, is suggestive.[67] With a short stature (170 cm) for a cross-country skier, Mäntyranta won a total of seven medals in long-distance events in three Winter Olympics (1960–1968). Yet, his achievements came along with suspicions of blood doping, owing to his very high hemoglobin concentration.

Twenty years after his retirement as a competitive athlete, Mäntyranta's blood was analyzed by the Finnish geneticist Albert de la Chapelle. The hemoglobin concentration was very high, 23.6 gr per dL, more than a 50% increment compared with normal values.[67] The genetic mutation underlying PFCP also affected 29 of

Mäntyranta's family members, two of whom also were Olympic medalists in cross-country skiing. Of note, there are multiple types of primary polycythemias, i.e., alterations leading to increased red blood cells, usually associated with adverse effects on health such as high blood pressure and elevated risk of blood clots and stroke. However, none of Mäntyranta's family members presented pathological alterations related to PFCP. De la Chapelle thus found a rare mutation highly beneficial for endurance capacity with few consequences for health, described as benign human erythrocytosis.[67] Mäntyranta's superb exercise capacity was not exclusively the result of his strong determination and psyche, as he stubbornly affirmed; but it was, at least in part, bestowed by Mother Nature. In any case, his impressive bronze statue, located in the small village of Pello, 20 km from the Arctic Circle, is *naturally* deserved.

The hormone erythropoietin can be manufactured, thus representing an extremely attractive drug to enhance endurance performance.[68] Extraordinary blood O_2 carrying capacity can be encountered in athletes doping with erythropoietin and/or blood transfusion.[69] In these, 20 gr of hemoglobin per dL of blood or greater has been reported in parallel to outstanding athletic achievements in endurance disciplines.[67, 70] In fact, such abnormally enhanced blood O_2 carrying capacity concurs with optimal values for endurance performance in experimental studies in mice.[71] In this regard, researchers at the Institute of Veterinary Physiology of Zürich developed genetically modified mice with constitutive overexpression of erythropoietin, resulting in extremely high blood O_2 carrying capacity (90% hematocrit, i.e., the ratio of total red blood cell volume to plasma volume), which was gradually reduced via pharmacological hemolysis.[71] VO_{2max} and endurance performance (time to exhaustion) were tested at progressively lower hematocrit levels, starting at 90% hematocrit. Following a curvilinear relationship, the greatest values for VO_{2max} and time to exhaustion were observed with hematocrits between 60% and 70%, equivalent to 20–23 gr of hemoglobin per dL of blood. Slightly lower optimal levels, ~58% hematocrit, were noted in normal mice with short-term elevation of blood O_2 carrying capacity via exogenous erythropoietin administration. Hence, the lifetime exposure to elevated blood O_2 carrying capacity facilitates further gains in genetically modified mice. Perhaps recurrent blood doping provides a similar *familiarization* effect in humans.

ENHANCEMENT OF O_2 CARRYING CAPACITY: THE CREATION OF A MYTH

Not only mice and humans benefit from enhanced blood O_2 carrying capacity. Aerobically 'athletic' animals such as the greyhound dog exhibit ~20 g of hemoglobin per dL of blood at rest, further increased during exercise partly due to red blood cell release into the circulation by spleen contraction.[72] Exercise-induced spleen contraction can be minimally elicited in humans and has little impact on circulating red blood cells, irrespective of fitness status.[73–75] Would too many red blood cells, i.e., high hematocrit, impair the circulation in humans? Well, the clinical issue of potentially increased blood viscosity in blood-doped athletes has been put forward for decades. Nothing less than 36 academic texts cited the following story with some variation in the dates and number of dead cyclists:

> Between 1987 and 1990, 18 [Dutch and Belgium] cyclists died tragically and suddenly all from heart attack or stroke. Erythropoietin was known to thicken the blood—the common cause of heart attack or stroke. Many victims developed clots that broke off and travelled to their hearts or brains; others died of simple cardiac arrest, the organ struggling to pump blood the consistency of oil.[50]

This sensational account is supposed to provide a major rationale for anti-doping efforts—i.e., preventing deaths of healthy young individuals. That notwithstanding, how much truth, if any, underlies this story? In 2011, the Spanish sociologist Bernat Lopez attempted to answer that question.[76] He searched for and reviewed available academic, newspaper and popular sources as well as clinical databases reporting sudden death of cyclists in Europe from 1987 to 2010. The majority of documents retrieved either failed to quote the source of their claim or quoted another source that failed to do so. Other documents referred to an expert or journalistic source that did not substantiate the claim. Among all deceased cyclists during the late 1980s and early 1990s in Europe, not a single one could be linked somehow to the use of erythropoietin. In fact, there are experimental data supporting the rather innocuous nature of erythropoietin in specific supraphysiological doses. Lopez concluded that 'existing truly experimental and epidemiological research downplays or even rules out the existence of a causal link between erythropoietin intake and sudden death in healthy adults'. As specified in title of his article, a tacit conspiracy emerged to fabricate a 'drug of mass destruction'.[76] For what purpose? Perhaps for anti-doping propaganda, as aforementioned, to justify the existence and public funding of certain 'caring' institutions in the world of sport. Or perhaps nothing more than a good-hearted, but uninformed, attempt to deter athletes from jeopardizing their health—an overwhelmingly failed attempt according to the analyses of blood samples from elite athletes in multiple countries.[52] As the Danish doping researcher Verner Møller recalls, 'the road to Hell is paved with good intentions'.[77]

Ultimately, it is worth noting that neither endurance training nor any other known lifestyle intervention enhances blood O_2 carrying capacity—i.e., hemoglobin concentration—in healthy humans.[78] In fact, such a major hematological limitation refutes, from a physiological viewpoint, the widespread notion that humans evolved with extraordinary potential for endurance exercise capacity.[79] Our species did not evolve with the urgency to excel at endurance 'events', in striking contrast with other non-primate mammals.[80, 81] Indeed, we are definitely *made* to maintain a fixed level of blood O_2 carrying capacity that is incompatible with outstanding aerobic exercise capacity.

THE INHERENT FEMALE HANDICAP IN BLOOD O_2 CARRYING CAPACITY

Red blood cells constitute a highly valued asset in biological terms. In every cubic millimeter of blood there are approximately 5 million red blood cells, 25 trillion in the whole body—by far the most abundant type of cells in humans, comprising above 80% of all cells in our body due to their minuscule size.[82] Besides being legion, red blood cells have a high turnover rate: they circulate for about 120 days until being phagocytized by macrophages (16, 194). Consequently, every day, the bone marrow in healthy adults must produce more than 200 billion red blood cells, comprising the protein hemoglobin, to maintain normal O_2 carrying capacity. Such a large-scale anabolic process requires diverting energy and providing scarce essential resources (i.e., iron) from other biological demands. In females, the prospect of pregnancy might be one of those major competing demands. A consistently lower number of circulating red blood cells and hemoglobin is indeed found in females compared with males in the vast majority of vertebrates and almost all mammal species studied to date.[83] Humans are no exception.

Healthy adult women generally present a 10–15 g deficit of hemoglobin per liter of blood compared with men matched by age and physical activity (Figure 3.1).[83, 84] The sex gap in hemoglobin concentration is maintained throughout the spectrum of aerobic capacities, from sedentary to elite endurance athletes.[85] In percentage terms, the female sex is constitutionally endowed with a ~10% decrement in blood O_2 carrying capacity. Experimental studies have recently demonstrated that a given reduction (10%) in blood O_2 carrying capacity, while keeping unaltered other key hematological features, results in equivalent decrements in VO_{2max} in both sexes.[84] Hence, hemoglobin concentration proportionally determines aerobic capacity, irrespective of female or male sex. The persistent hematological *handicap* implies that women carry on any sustainable absolute workload at a higher fraction of aerobic capacity and thus augmented cardiovascular and hemodynamic stress than men. For instance, cycling or running at a certain absolute intensity (power output or velocity) must require higher cardiac output in women, since their blood transports less O_2 per unit of volume. Importantly, the prevalence of anemia, i.e., low blood hemoglobin concentration, increases with advanced age even in the absence of underlying pathological conditions.[86] Older women thus present the lowest capacity to perform aerobic exercise among healthy adults. Not surprisingly, they comprise the fraction of the population most likely to be affected by severe cardiac disabilities primarily associated with low physical activity.[87]

The question remains as to why the female sex presents lower blood O_2 carrying capacity. As it must be already clear, there is no physiological limitation to enhance erythropoiesis in humans. In fact, women can increase their hemoglobin concentration in response to erythropoietin to the same magnitude as men.[88] Likewise, women living at high altitude (>4000 m) present higher erythropoietin and hemoglobin concentration than women at sea level, while the sex gap compared with altitude-living men is preserved.[89] Females' physiology thus *chooses* to have a lower blood O_2 carrying capacity. In fact, the intrinsic female scarcity of androgenic stimuli (e.g., testosterone) for red blood cell formation could be compensated by increased erythropoietin production, but it is not. Low O_2 in females' blood may be related to some aspect of evolution with a forthright narrative—it would be unlikely (yet possible) for a condition such as sex dimorphism to be so prevalent and have independently emerged (e.g., in birds and mammals) by chance. Is the 'nature' of females to be less physically active, thus requiring less aerobic capacity, than males? Perhaps humans evolved in this direction, but this was not necessarily so for all mammal species. For instance, rodent studies suggest that under normal conditions, females are more physically active than males, which has been linked to food acquisition for their offspring,[90] although there are exceptions to such a generalization.[91, 92] Until more evidence becomes available, we may speculate that the inescapable trade-off between biological cost and benefit of 'expensive' biological variables likely contributed to a lower *need* for blood O_2 carrying capacity in women. They are the ones who must provide the essential nutrients in the most critical stages of human development.

PARALLELISM BETWEEN SEX GAPS IN O_2 DELIVERY AND ELITE ENDURANCE PERFORMANCE

The impact of blood O_2 carrying capacity on endurance performance becomes prominent from submaximal (~80% HR_{max}) up to maximal aerobic exercise capacity (occurring at ~100% HR_{max}), a range of intensities in which O_2 delivery has a major impact on O_2 uptake.[93] Consequently, in the presence of normal lung function, blood (arterial) O_2 content determines aerobic energy production at the exercise intensities at which most endurance competitions are carried out. Hemoglobin concentration must therefore contribute to sex differences in endurance performance. How large is the sex gap in endurance performance? Looking at current world records from middle-distance to long-distance running (right after the Tokyo 2020 Olympic Games), women present 10–12% longer times relative to men—a quantitatively similar percentage difference to that in hemoglobin concentration between sexes. The consistency in the performance gap across distances that requires high aerobic capacity is striking: 1500 m (12% longer world record in women), 3000 m (10%), 5000 m (12%), 10000 m (11%) and marathon (10%).

The extent to which such differences in performance would be attenuated if hemoglobin concentration were unisex remains to be elucidated.[37] What can be assumed with some confidence is that blood O_2 carrying capacity plausibly explains a substantial portion of the sex-specific variance in endurance performance.[4, 84]

A question may arise at this point: if women have a 10% lower blood O_2 carrying capacity and, as mentioned in the previous chapter, maximal cardiac output per body weight in female elite endurance athletes is ~15% lower than that of male counterparts,[4] the percentage sex difference in VO_{2max} should be larger (~25%) than that noted for records in endurance performance.[4] In this regard, the first point to acknowledge is that VO_{2max} and endurance performance are strongly correlated, but they are not fully determined by the same cluster of factors. Two potential explanations respectively related to physiology and pharmacology may underlie the discordance between sex differences in VO_{2max} and performance. Physiologically, skeletal muscle attributes that determine fuel utilization differ between sexes,[37] plausibly favoring females—as detailed in the next chapter on skeletal muscle features. That notwithstanding, the contribution of strictly peripheral traits to endurance performance should be minor in healthy individuals.[4, 94] Indeed, no sex differences in the percentage of VO_{2max} that can be sustained during endurance events, which is presumed to be partly determined by peripheral factors, have been established.[95] Hence, the main focus must be placed on the alternative potential explanation: pharmacology, more specifically on doping.

Any substance or procedure that is effective in improving performance will always be a strong temptation for elite athletes and their entourage. Blood doping is highly effective, whereas anti-doping efforts are largely unsucessful.[68] In this scenario, the International Cycling Union established the 'no start' rule in 1997, which was later adopted by other endurance sports federations. According to this rule, any athlete with a blood O_2 carrying capacity above a fixed threshold was deemed 'unfit' and prevented from competition for 15 days from the date of the test. The threshold was set according to the hematocrit: 50% (~16.7 gr of hemoglobin per dL of blood) and 47% (~15.7 gr of hemoglobin per dL of blood) for men and women, respectively. Predictably, after this legislation was settled, the hematocrit of most elite cyclists during competition was just slightly below the threshold.[96] Yet, due to the omnipresent biological variability, a minority of athletes present naturally elevated blood O_2 carrying capacity levels above the threshold—these athletes can compete following a tedious appeal including medical certification of their abnormal hematological values.[85] In this regard, cross-sectional investigations have found a substantially larger fraction of male relative to female elite endurance athletes beyond the threshold, seemingly in the absence of doping.[85] This suggests that the *hematological reserve* for artificially increasing blood O_2 carrying capacity and still remain in the legal range is wider in women. A greater performance enhancement via blood doping may thus be available to women. The historically intimate relationship of world athletic records with doping makes this possibility worthy of consideration. Alternatively, a relatively larger fraction of the female athletic population with naturally extreme levels of blood O_2 carrying capacity can freely compete (without the justification process required for men), which can also contribute to enhance performance achievements.[85] That notwithstanding, the notion that athletic feats are performed only by innately gifted (biologically) individuals has been frequently refuted.[50, 96, 97] It should not be forgotten that elite athletes, women and men, are outliers in essence, comprising an exorbitant will to succeed.

REFERENCES

1. Diaz-Canestro C, Pentz B, Sehgal A, Montero D. Sex differences in cardiorespiratory fitness are explained by blood volume and oxygen carrying capacity. *Cardiovasc Res.* 2021;118(1):334–343.
2. Hellsten Y, Nyberg M. Cardiovascular adaptations to exercise training. *Compr Physiol* 2015;6(1):1–32.
3. Montero D, Lundby C. Red cell volume response to exercise training: Association with aging. *Scand J Med Sci Sports* 2017;27(7):674–683.
4. Lundby C, Robach P. Performance enhancement: What are the physiological limits? *Physiology (Bethesda)* 2015;30(4):282–292.
5. Kjellberg SR, Rudhe U, Sjöstrand T. Increase of the amount of hemoglobin and blood volume in connection with physical training. *Acta Physiol Scand* 1949;19(2–3):146–151.
6. Dill DB, Braithwaite K, Adams WC, Bernauer EM. Blood volume of middle-distance runners: Effect of 2,300-m altitude and comparison with non-athletes. *Med Sci Sports* 1974;6(1):1–7.
7. Brotherhood J, Brozovic B, Pugh LG. Haematological status of middle- and long-distance runners. *Clin Sci Mol Med* 1975;48(2):139–145.
8. Heinicke K, Wolfarth B, Winchenbach P, et al. Blood volume and hemoglobin mass in elite athletes of different disciplines. *Int J Sports Med* 2001;22(7):504–512.

9. Karpovich PV, Millman N. Athletes as blood donors. *Res Q Am Assoc Health Phys Educ Recreat* 1941;13(2):166–168.
10. Howell ML, Coupe K. Effect of blood loss upon performance in the balke-ware treadmill test. *Res Q* 1964;35:156–165.
11. Hollmann W, Chirdel K, Forsberg S, Speer K. Studies on the effect of blood donation on cardiopulmonary performance. *Med Welt* 1969;20:1158–1161.
12. Ekblom B, Goldbarg AN, Gullbring B. Response to exercise after blood loss and reinfusion. *J Appl Physiol* 1972;33(2):175–180.
13. Williams MH, Lindhjem M, Schuster R. The effect of blood infusion upon endurance capacity and ratings of perceived exertion. *Med Sci Sports* 1978;10(2):113–118.
14. Markiewicz K, Cholewa M, Gorski L, Jaszczuk J, Chmura J, Bartniczak Z. Effect of 400 ml blood loss on adaptation of certain functions of the organism to exercise. *Acta Physiol Pol* 1981;32(6):613–621.
15. Fritsch J, Winter UJ, Reupke I, Gitt AK, Berge PG, Hilger HH. Effect of a single blood donation on ergo-spirometrically determined cardiopulmonary performance capacity of young healthy probands. *Z Kardiol* 1993;82(7):425–431.
16. Krip B, Gledhill N, Jamnik V, Warburton D. Effect of alterations in blood volume on cardiac function during maximal exercise. *Med Sci Sports Exerc* 1997;29(11):1469–1476.
17. Janetzko K, Bocher R, Klotz KF, Kirchner H, Kluter H. Effects of blood donation on the physical fitness and hemorheology of healthy elderly donors. *Vox Sang* 1998;75(1):7–11.
18. Duda K, Zoladz JA, Majerczak J, Kolodziejski L, Konturek SJ. The effect of exercise performed before and 24 hours after blood withdrawal on serum erythropoietin and growth hormone concentrations in humans. *Int J Sports Med* 2003;24(5):326–331.
19. Birnbaum L, Dahl T, Boone T. Effect of blood donation on maximal oxygen consumption. *J Sports Med Phys Fitness* 2006;46(4):535–539.
20. Burnley M, Roberts CL, Thatcher R, Doust JH, Jones AM. Influence of blood donation on O_2 uptake on-kinetics, peak O_2 uptake and time to exhaustion during severe-intensity cycle exercise in humans. *Exp Physiol* 2006;91(3):499–509.
21. Dellweg D, Siemon K, Mahler F, Appelhans P, Klauke M, Kohler D. Cardiopulmonary exercise testing before and after blood donation. *Pneumologie* 2008;62:372–377.
22. Foster C, Porcari JP, Anderson J, et al. The talk test as a marker of exercise training intensity. *J Cardiopulm Rehabil Prev* 2008;28(1):24–30; quiz 31–22.
23. Gordon D, Marshall K, Connell A, Barnes RJ. Influence of blood donation on oxygen uptake kinetics during moderate and heavy intensity cycle exercise. *Int J Sports Med* 2010;31(5):298–303.

24. Judd TB, Cornish SM, Barss TS, Oroz I, Chilibeck PD. Time course for recovery of peak aerobic power after blood donation. *J Strength Cond Res* 2011;25(11):3035–3038.
25. Mora-Rodriguez R, Aguado-Jimenez R, Del Coso J, Estevez E. A standard blood bank donation alters the thermal and cardiovascular responses during subsequent exercise. *Transfusion* 2012;52(11):2339–2347.
26. Hill DW, Vingren JL, Burdette SD. Effect of plasma donation and blood donation on aerobic and anaerobic responses in exhaustive, severe-intensity exercise. *Appl Physiol Nutr Metab* 2013;38(5):551–557.
27. Strandenes G, Skogrand H, Spinella PC, Hervig T, Rein EB. Donor performance of combat readiness skills of special forces soldiers are maintained immediately after whole blood donation: A study to support the development of a prehospital fresh whole blood transfusion program. *Transfusion* 2013;53(3):526–530.
28. Gordon D, Wood M, Porter A, et al. Influence of blood donation on the incidence of plateau at VO2max. *Eur J Appl Physiol* 2014;114(1):21–27.
29. Ziegler AK, Grand J, Stangerup I, et al. Time course for the recovery of physical performance, blood hemoglobin, and ferritin content after blood donation. *Transfusion* 2015;55(4):898–905.
30. Meurrens J, Steiner T, Ponette J, et al. Effect of repeated whole blood donations on aerobic capacity and hemoglobin mass in moderately trained male subjects: A randomized controlled trial. *Sports Med Open* 2016;2(1):43.
31. Panebianco RA, Stachenfeld N, Coplan NL, Gleim GW. Effects of blood donation on exercise performance in competitive cyclists. *Am Heart J* 1995;130(4):838–840.
32. Diaz Canestro C, Pentz B, Sehgal A, Montero D. Blood withdrawal acutely impairs cardiac filling, output and aerobic capacity in proportion to induced hypovolemia in middle-aged and older women. *Appl Physiol Nutr Metab*. 2021;47(1):75–82.
33. American College of Sport Medicine (ACSM). *ACSM's Guidelines for Exercise Testing and Prescription*, 10th Edition. Philadelphia, PA: LWW; 2017.
34. Munoz M, Acheson AG, Bisbe E, et al. An international consensus statement on the management of postoperative anaemia after major surgical procedures. *Anaesthesia* 2018;73(11):1418–1431.
35. Diaz-Canestro C, Pentz B, Sehgal A, Montero D. Differences in cardiac output and aerobic capacity between sexes are explained by blood volume and oxygen carrying capacity. *Front Physiol* 2022;13:747903.
36. Montero D, Madsen K, Meinild-Lundby AK, Edin F, Lundby C. Sexual dimorphism of substrate utilization: Differences in skeletal muscle mitochondrial volume density and function. *Exp Physiol* 2018;103(6):851–859.

37. Diaz-Canestro C, Montero D. Unveiling women's powerhouse. *Exp Physiol* 2020;105(7):1060–1062.
38. Sheel AW, Dominelli PB, Molgat-Seon Y. Revisiting dysanapsis: Sex-based differences in airways and the mechanics of breathing during exercise. *Exp Physiol* 2016;101(2):213–218.
39. Montero D, Lundby C. Regulation of red blood cell volume with exercise training. *Compr Physiol* 2018;9(1):149–164.
40. Lundby C, Robach P, Saltin B. The evolving science of detection of 'blood doping'. *Br J Pharmacol* 2012;165(5):1306–1315.
41. Patterson SW, Starling EH. On the mechanical factors which determine the output of the ventricles. *J Physiol* 1914;48(5):357–379.
42. Bundgaard-Nielsen M, Sørensen H, Dalsgaard M, Rasmussen P, Secher NH. Relationship between stroke volume, cardiac output and filling of the heart during tilt. *Acta Anaesthesiol Scand* 2009;53(10):1324–1328.
43. Gregersen M, Rawson RA. Blood volume. *Physiol Rev* 1959(39):307–342.
44. Montero D, Cathomen A, Jacobs RA, et al. Haematological rather than skeletal muscle adaptations contribute to the increase in peak oxygen uptake induced by moderate endurance training. *J Physiol* 2015;593(20):4677–4688.
45. Munch GD, Svendsen JH, Damsgaard R, Secher NH, Gonzalez-Alonso J, Mortensen SP. Maximal heart rate does not limit cardiovascular capacity in healthy humans: Insight from right atrial pacing during maximal exercise. *J Physiol* 2014;592(2):377–390.
46. Montero D, Lundby C, Ruschitzka F, Flammer AJ. True anemia-red blood cell volume deficit-in heart failure: A systematic review. *Circ Heart Fail.* 2017;10(5):e003610.
47. Seymour WB, Pritchard WH, Longley LP, Hayman JM. Cardiac output, blood and interstitial fluid volumes, total circulating serum protein, and kidney function during cardiac failure and after improvement. *J Clin Investig* 1942;21(2):229–240.
48. Kaplan E, Puestow RC, Baker LA, Kruger S. Blood volume in congestive heart failure as determined with iodinated human serum albumin. *Am Heart J* 1954;47(5):824–838.
49. Watson RD, Gibbs CR, Lip GY. ABC of heart failure. Clinical features and complications. *BMJ* 2000;320(7229):236–239.
50. Parisotto R. *Blood Sports: The Inside Dope on Drugs in Sport.* Self; 2010.
51. Solheim SA, Bejder J, Breenfeldt Andersen A, Morkeberg J, Nordsborg NB. Autologous blood transfusion enhances exercise performance-strength of the evidence and physiological mechanisms. *Sports Med Open* 2019;5(1):30.
52. Sottas PE, Robinson N, Saugy M. The athlete's biological passport and indirect markers of blood doping. *Handb Exp Pharmacol* 2010(195):305–326.

53. Faiss R, Saugy J, Zollinger A, et al. Prevalence estimate of blood doping in elite track and field athletes during two major international events. *Front Physiol.* 2020;11:160.
54. Ulrich R, Pope HG, Jr., Cleret L, et al. Doping in two elite athletics competitions assessed by randomized-response surveys. *Sports Med* 2018;48(1):211–219.
55. Malm CB, Khoo NS, Granlund I, Lindstedt E, Hult A. Autologous doping with cryopreserved red blood cells - effects on physical performance and detection by multivariate statistics. *PLOS ONE* 2016;11(6):e0156157.
56. Sorensen EA. Debunking the myth of pregnancy doping. *J Intercollegiate Sport* 2009;2:269–285.
57. Paulev PE. *Textbook in Medical Physiology and Pathophysiology, Essentials and Clinical Problems.* 1999.
58. Wolff A, O'Brien R. Pregnancy doping. *Sports Illustrated* 1994;81(23):16.
59. https://www.si.com/olympics/2017/06/06/sanya-richards-ross-opens-about-abortion. 2017.
60. Pritchard JA. Changes in the blood volume during pregnancy and delivery. *Anesthesiology* 1965;26:393–399.
61. Kjellberg SR, Lonroth H, Rudhe U, Sjostrand T. Blood volume and heart volume during pregnancy and the puerperium. *Acta Med Scand* 1950;138(6):421–429.
62. Bonne TC, Doucende G, Fluck D, et al. Phlebotomy eliminates the maximal cardiac output response to six weeks of exercise training. *Am J Physiol Regul Integr Comp Physiol* 2014;306(10):R752–R760.
63. Mandic M, Eriksson LMJ, Melin M, et al. Increased maximal oxygen uptake after sprint-interval training is mediated by central haemodynamic factors as determined by right heart catheterization. *J Physiol* 2023;601(12):2359–2370.
64. Montero D, Lundby C. Refuting the myth of non-response to exercise training: 'non-responders' do respond to higher dose of training. *J Physiol* 2017;595(11):3377–3387.
65. Ekblom B, Astrand PO, Saltin B, Stenberg J, Wallstrom B. Effect of training on circulatory response to exercise. *J Appl Physiol* 1968;24(4):518–528.
66. Tanaka H, Bassett DR, Jr., Turner MJ. Exaggerated blood pressure response to maximal exercise in endurance-trained individuals. *Am J Hypertens* 1996;9(11):1099–1103.
67. de la Chapelle A, Traskelin AL, Juvonen E. Truncated erythropoietin receptor causes dominantly inherited benign human erythrocytosis. *Proc Natl Acad Sci U S A* 1993;90(10):4495–4499.
68. Thomsen JJ, Rentsch RL, Robach P, et al. Prolonged administration of recombinant human erythropoietin increases submaximal performance more than maximal aerobic capacity. *Eur J Appl Physiol* 2007;101(4):481–486.

69. USADA. *Report on proceedings under the world anti-doping code and the USADA protocol United States anti-doping agency, claimant, v. Lance Armstrong, respondent. Reasoned decision of the USADA.* 2012.
70. Hamilton T, Coyle D. *The secret race. Inside the hidden world of the Tour de France.* New York: Bantam (reprint ed.);2013.
71. Schuler B, Arras M, Keller S, et al. Optimal hematocrit for maximal exercise performance in acute and chronic erythropoietin-treated mice. *Proc Natl Acad Sci U S A* 2010;107(1):419–423.
72. Uhrikova I, Lacnakova A, Tandlerova K, et al. Haematological and biochemical variations among eight sighthound breeds. *Aust Vet J* 2013;91(11):452–459.
73. Prommer N, Ehrmann U, Schmidt W, Steinacker JM, Radermacher P, Muth CM. Total haemoglobin mass and spleen contraction: A study on competitive apnea divers, non-diving athletes and untrained control subjects. *Eur J Appl Physiol* 2007;101(6):753–759.
74. Shephard RJ. Responses of the human spleen to exercise. *J Sports Sci* 2016;34(10):929–936.
75. Engan HK, Lodin-Sundstrom A, Schagatay F, Schagatay E. The effect of climbing Mount Everest on spleen contraction and increase in hemoglobin concentration during breath holding and exercise. *High Alt Med Biol* 2014;15(1):52–57.
76. Lopez B. The invention of a 'drug of mass destruction': Deconstructing the EPO myth. *Sport in History.* 2011;31(1):84–109.
77. Moller V. *The Ethics of Doping and Anti-doping.* 2010.
78. Lundby C, Montero D. Did you know-why does maximal oxygen uptake increase in humans following endurance exercise training? *Acta Physiol (Oxf)* 2019;227(4):e13371.
79. Pickering TR, Bunn HT. The endurance running hypothesis and hunting and scavenging in savanna-woodlands. *J Hum Evol* 2007;53(4):434–438.
80. Knight PK, Sinha AK, Rose RJ. *Effects of training intensity on maximum oxygen uptake.* Davis, CA: ICEEP Publications; 1991.
81. Musch TI, Haidet GC, Ordway GA, Longhurst JC, Mitchell JH. Dynamic exercise training in foxhounds. I. Oxygen consumption and hemodynamic responses. *J Appl Physiol* 1985;59(1):183–189.
82. Sender R, Fuchs S, Milo R. Revised estimates for the number of human and bacteria cells in the body. *PLoS Biol* 2016;14(8):e1002533.
83. Murphy WG. The sex difference in haemoglobin levels in adults—mechanisms, causes, and consequences. *Blood Rev* 2014;28(2):41–47.
84. Diaz-Canestro C, Siebenmann C, Montero D. Blood oxygen carrying capacity determines cardiorespiratory fitness in middle-age and older women and men. *Med Sci Sports Exerc.* 2021;53(11):2274–2282.
85. Schumacher YO, Jankovits R, Bultermann D, Schmid A, Berg A. Hematological indices in elite cyclists. *Scand J Med Sci Sports* 2002;12(5):301–308.

86. Montero D, Diaz-Canestro C, Flammer A, Lundby C. Unexplained anemia in the elderly: Potential role of arterial stiffness. *Front Physiol* 2016;7:485.
87. Diaz-Canestro C, Montero D. Female sex-specific curtailment of left ventricular volume and mass in HFpEF patients with high end-diastolic filling pressure. *J Hum Hypertens* 2020.
88. Morkeberg J, Lundby C, Nissen-Lie G, Nielsen TK, Hemmersbach P, Damsgaard R. Detection of darbepoetin alfa misuse in urine and blood: A preliminary investigation. *Med Sci Sports Exerc* 2007;39(10):1742–1747.
89. Beall CM, Brittenham GM, Strohl KP, et al. Hemoglobin concentration of high-altitude Tibetans and Bolivian Aymara. *Am J Phys Anthropol* 1998;106(3):385–400.
90. Rosenfeld CS. Sex-dependent differences in voluntary physical activity. *J Neurosci Res* 2017;95(1–2):279–290.
91. Kane JD, Steinbach TJ, Sturdivant RX, Burks RE. Sex-associated effects on hematologic and serum chemistry analytes in sand rats (Psammomys obesus). *J Am Assoc Lab Anim Sci* 2012;51(6):769–774.
92. Gromov VS. Daytime activities and social interactions in a colony of the fat sand rats, Psammomys obesus, at the Negev Highlands, Israel. *Mammalia* 2001;65(1):13–21.
93. Mortensen SP, Damsgaard R, Dawson EA, Secher NH, Gonzalez-Alonso J. Restrictions in systemic and locomotor skeletal muscle perfusion, oxygen supply and VO2 during high-intensity whole-body exercise in humans. *J Physiol* 2008;586(10):2621–2635.
94. Lundby C, Montero D, Joyner M. Biology of VO2 max: Looking under the physiology lamp. *Acta Physiol (Oxf)* 2017;220(2):218–228.
95. Helgerud J. Maximal oxygen uptake, anaerobic threshold and running economy in women and men with similar performances level in marathons. *Eur J Appl Physiol Occup Physiol* 1994;68(2):155–161.
96. Di Luca D. *Cycliste infiltré*. CITY;2017.
97. Millar D. *Racing Through the Dark: The Fall and Rise of David Millar*. Orion; 2011.

Chapter 4

Skeletal muscle

Abbreviations:

ATP, adenosine triphosphate
DNA, deoxyribonucleic acid
VO_{2max}, maximal oxygen consumption

HUMANS: MUSCULAR VERSATILE ANIMALS

Skeletal muscle has attracted the attention of several generations of exercise physiologists. Much of our understanding of muscle physiology derives from the application of surgical procedures to obtain muscle biopsies, the most commonly used being the Bergström technique. Hereby, a biopsy needle is inserted into a large skeletal muscle, e.g., the vastus lateralis. The needle has a surgical blade inside that cuts a small portion of muscle, which remains inside the needle. Approximately 150 mg of skeletal muscle is taken in a muscle biopsy, which represents a minor fraction of the vastus lateralis.[1] This piece of tissue, similar in size to a pencil eraser, provides a wealth of information on the structure and function of skeletal muscle at the cellular and molecular levels. Whether the tissue removed by the biopsy is mainly regenerated in the form of skeletal muscle or non-functional connective tissue remains uncertain.[2] Knowledge usually has a price. Paradoxically, the fact that some 'veteran' researchers in muscle physiology accumulate numerous muscle biopsies in their own vastus lateralis and still maintain normal muscle function might imply the *massive* irrelevance of their object of study. Surely not—besides, relevance is a relative and subjective concept.

DOI: 10.1201/9781003486893-4

Men and women of knowledge generally have a strong tendency to classify. From the advent of the characterization of skeletal muscle fibers, they have been assorted in categories, despite the fact that there is a continuous spectrum of functional and biochemical muscle fiber characteristics. The main differentiation is made between type 1 (slow-twitch) and type 2 (fast-twitch) muscle fibers. Type 1 contract more slowly, generate less peak force per contraction at a lower energy cost (per unit of force) and resist longer until failure (fatigue) than type 2 muscle fibers. Type 1 are thus more suitable for prolonged, endurance efforts, whereas type 2 are convenient for efforts requiring high levels of strength and power. Endurance and strength/power athletes typically have a greater proportion of type 1 and type 2 fibers in their trained muscles, respectively, than sedentary individuals.[3] Yet, human skeletal muscles are not 'single-fibered'. In contrast to other mammals, humans commonly have an ample mix of skeletal muscle fibers in each muscle.[4] No human muscle is thus extremely specialized for a specific set of contractile characteristics. Muscularly speaking, we are versatile, multifaceted animals that do not excel in any particular physical activity, as compared with *athletic* mammals.[5, 6]

SKELETAL MUSCLE FIBER TYPE AND EXERCISE CAPACITY

In previous chapters we have underlined the crucial role of O_2 delivery via the cardiovascular system for VO_{2max} and thereby endurance performance. Briefly stated, it is impossible to reach high levels of VO_{2max} without a correspondingly high capacity to deliver O_2 to the tissues.[7] On the other hand, could it be possible to excel in aerobic capacity and endurance performance without a high proportion of type 1 fibers? According to the evidence provided by skeletal muscle biopsies in elite distance runners, muscle fiber type is a poor predictor of marathon performance: similarly outstanding racing times can be achieved with a very large range of type 1 fibers (27% to 98%) in active running muscles.[8] Does this mean that muscle fiber type is irrelevant for exercise (endurance) capacity? Not necessarily, as developed in the following paragraph.

Skeletal muscles are not all-or-none contractile units. Each muscle is comprised of hundreds of motor neurons, each innervating hundreds to thousands of muscle fibers. All motor neurons and their muscle fibers are not activated during exhausting exercise, as determined from the total energetic substrate remaining in the muscle. Even at the point of muscular failure in an incremental cycling exercise test, a substantial reserve of fast energy-releasing substrates (adenosine triphosphate (ATP), phosphocreatine) are left unused (not recruited) in main quadriceps muscles.[9] Likewise, glycogen reserves in type 1 and type 2 fibers are far from being depleted at exhaustion with either lighter or heavier resistance loads.[10] Muscular failure is thus reached when a fraction of muscle fibers still have the intrinsic potential to contract, a puzzling fact. The underlying explanation—i.e., the primary determinants of muscular failure at the molecular, muscle and integrated system levels—remains uncertain.[9] Yet, a clear inference can be made: the activation (recruitment) of muscle fibers, in addition to or rather than the proportion of muscle fiber types, cannot be neglected.

In the previous example, it is possible that marathon runners with a low percentage of type 1 fibers activate more of them (and thus fewer type 2 fibers) than those with a high percentage, so that both end up with a similar total number of activated type 1 fibers. Thus far, conclusive data from studies comparing muscle activation in similarly fit individuals presenting with large differences in the percentage of muscle fiber type has not been reported or perhaps even investigated. In fact, straightforward measurements of how many and what type of muscle fibers are activated in a given muscular contraction in exercising humans still involve major technical challenges. If feasible, the interpretation of these measurements would be necessarily compounded by the assumption that a small superficial fraction of muscle analyzed in a biopsy can accurately reflect the composition of the whole muscle.[11] That notwithstanding, on average, marathon runners and endurance athletes in general have a higher percentage of type 1 fibers and a lower percentage of type 2 fibers than strength/power athletes.[12] This observation suggests some degree of association of muscle fiber type with exercise capacity, partly independent of muscle fiber activation. Nonetheless, in the absence of experimental evidence in humans, muscle fiber activation cannot be excluded as a potential primary factor explaining exercise capacity, overriding muscle fiber composition. The latter could be the *consequence* of an innate specific pattern of muscle activation.[13] We certainly know less than we seem to know according to textbooks.

A NOTE ABOUT SKELETAL MUSCLE FIBER TYPE BOUNDARIES

Touching upon textbooks, the categorization of muscle fibers may be useful for simplification in introductory lessons in physiology. Yet, we should bear in mind that multiple histochemical, metabolic and functional categories of muscle fibers can be considered, which are not necessarily equivalent.[14] Namely, within a definite muscle fiber type according to the histochemical category, there can be a continuum spectrum of metabolic and functional characteristics. In fact, the molecular machinery for aerobic and anaerobic metabolism is generally large enough in type 1 and type 2 fibers to accommodate and adapt to a wide range of exercise stimuli. Aside from pleasant all-fitting narratives, the boundaries among muscle fiber types are not clear-cut, and substantial metabolic and functional overlap are present between muscle fibers purportedly belonging to different categories.

MUSCLE GROWTH AND TRANSFORMATION WITH EXERCISE TRAINING

We are born with a definite number and proportion of muscle fiber types, according to the category used. The number of muscle fibers in a given individual does not seem to change from early childhood into middle adulthood.[15, 16] Likewise, exercise training is not presumed to stimulate the formation of new muscle fibers, known as hyperplasia, at least to a relevant extent in humans[17]—a different conclusion can be made in other mammals, such as cats.[18] Indeed, the number of muscle fibers in the bulging biceps brachii of elite bodybuilders does not differ, on average, from that of untrained controls: it is the increased cross-sectional area of each muscle fiber that explains the hypertrophy.[17] Yet, bodybuilders with the largest biceps tend to have a higher number of muscle fibers, although there is a large interindividual variability in the number of muscle fibers, independently of training status, with some highly trained bodybuilders presenting lower muscle fibers than untrained controls. Collectively considered, the genetic endowment might partly determine the potential for muscle growth by setting the number of muscle fibers.

Unlike cats, humans do not adapt to exercise training by creating new muscle fibers, a process known as hyperplasia. But, can the muscle fiber type be 'metamorphosed' in response to exercise stimuli? A certain level of muscle transformation (prominent adaptation would be a more precise term) between subtypes of type 2 fibers has been demonstrated with relatively short-term (~3–6 months) endurance or strength training.[19] Moreover, some evidence indicates that short-term endurance training may induce a moderate shift of type 2 towards type 1 fibers.[19] The notion remains that larger shifts from type 2 to type 1 fibers might be induced with long-term (as in years) endurance training, albeit the longitudinal evidence is lacking. Nonetheless, a case study in middle-age (52-year-old) monozygotic twins, hence having the same genes and likely a very similar muscle fiber composition at birth, provided compelling support for muscle fiber 'metamorphosis'.[20] One twin was mostly sedentary during his adult life, while the other reported more than 30 years of endurance training. A very large proportion (95%) of muscle fibers in the biopsy of the vastus lateralis in the trained twin were type 1, whereas less than 40% were type 1 fibers in the untrained twin. This case study suggests the large potential of type 2 muscle fibers to shift from type 1 to type 2, i.e., from fast-twitch but fatigable phenotype to slow-twitch and fatigue-resistant one, with chronic endurance training.

The evidence provided in the previous paragraph concurs with the common observation that athletes from multiple disciplines tend to gain endurance but lose speed and power from the beginning to the end of their careers. Similarly obvious is the fact that endurance athletes generally extend their careers substantially longer (> 40 years of age) than strength and/or power athletes. Furthermore, a typical perception in the general population is that the performance of physical power (explosive)-related activities, partly relying on type 2 muscle fibers and neural factors, wane at a relatively greater rate than endurance capacity during the adult lifetime.[21] The 'human candle' of explosive wax seems to burn the fastest.

SEX DIFFERENCES IN SKELETAL MUSCLE: IS IT ALL ABOUT SIZE?

All measurable characteristics of an organism, except its genes, constitute the phenotype. The majority of variables discussed in this book belong to the phenotype. They are the product of the interaction between genes (genotype) and environment. Two individuals may have an identical genotype (e.g., monozygotic twins) but a different phenotype due to being exposed to distinct environments. In this context, environment is a broad concept encompassing all external (e.g., climate, food availability, sociocultural) and internal (physical and chemical stimuli within the body) conditions that act upon an organism. For instance, a lifetime exposure to endurance training can be a highly influential environmental factor. Conversely, an identical environment may result in a distinct phenotype in individuals with a diverse genotype. Women and men can never have the same genotype. It is precisely the genetic code which dictates the sex of an individual.

Women and men share the majority of the genetic code. From a total of ~20,000 genes comprising the human genome, ~800 are sex-specific, either because they are present only in men or epigenetically regulated in a different manner in each sex.[22] As aforementioned, the interaction of these genetic divergences along with the environment determine phenotypic differences between sexes. Of note, even if two individuals exhibit multiple genetic disparities, provided an identical environment, their phenotype may not differ. In fact, biological redundancy is not uncommon at molecular levels.[23] Distinct genes or combination thereof may result in similar outcomes.

As with any other organ or tissue in the body, similitudes and divergences in skeletal muscles of women and men have been documented. To sift and discern which difference is functionally relevant and to what extent is not an easy task; it requires deep understanding. Despite limited data, the notion has prevailed that for a given exercise training status, women have smaller muscle fiber cross-sectional areas (Figure 4.1), which mainly explain the reduced cross-sectional area of the whole muscle, resulting in limited strength relative to men.[24–26] Female muscle fibers seem to have less potential to grow and/or less hormonal (anabolic) stimulation to do so. In addition, a slightly lower number of muscle fibers per muscle might also contribute to smaller muscle cross-sectional area and lower potential for muscular hypertrophy and strength development in women.[24–26]

More controversial is whether a sex difference exists regarding the percentage of muscle fiber types. Until relatively recently, it was generally considered that women had a higher percentage of type 1 and a lower percentage of type 2 fibers than men.[12] Such a tenet fitted well with preconceptions linking lower explosive physical

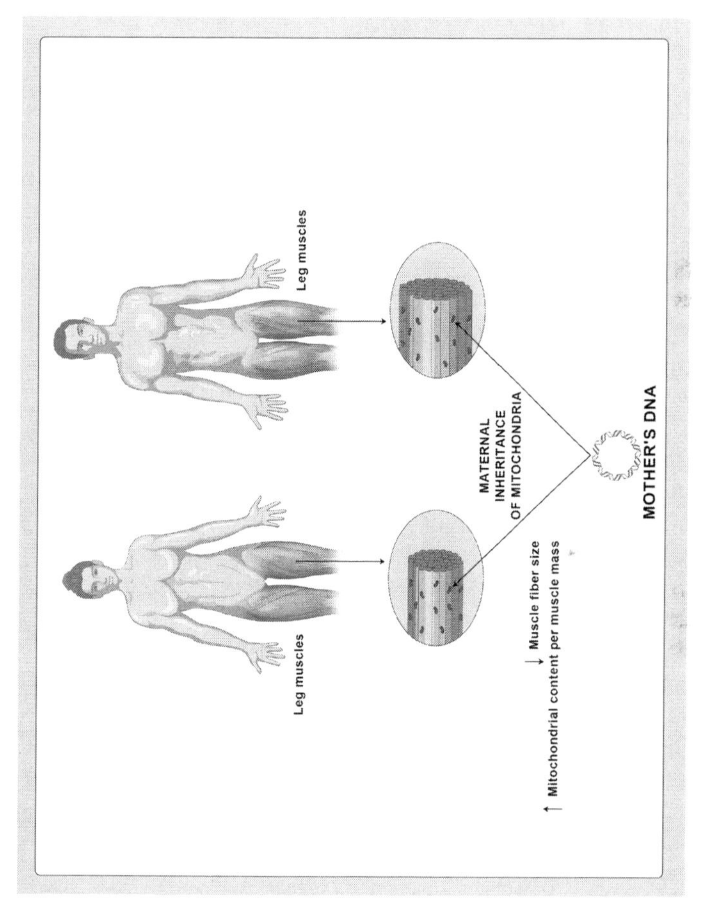

Figure 4.1 Sex differences in skeletal muscle fiber size and oxidative capacity. The man and woman in the figure have the same body weight.

capacity with the female sex. Yet, recent investigations have found similarly high (~70%) type 2 fiber percentage in elite women and men weightlifters.[27] Years of resistance training sustaining high weightlifting performance, rather than sex, seemed to explain such a high percentage of type 2 fibers. The authors stated: 'when you look at muscle tissue, you can't really differentiate between a man's muscle fibers and a woman's'.[28]

SEX PHENOTYPIC DIFFERENCES IN SKELETAL MUSCLE CONTENT AND FUNCTION

In relation with the final assertion of the previous paragraph, it should be noted that in the midst of the ruthless competition to publish, it is not uncommon for researchers to overstate their negative or positive findings, depending on their 'battle side'.[29, 30] Keeping this in mind, recent research has revealed a sex dimorphism in skeletal muscle.[31, 32] For the sake of clarity, the concept of sex dimorphism entails intrinsic phenotypic differences between males and females beyond reproductive organs. Increased mitochondrial content in skeletal muscle fibers has been recently found in healthy women compared with men matched by aerobic capacity (Figure 4.1).[31] Mitochondria in skeletal muscle fibers are the terminal structure in the O_2 cascade (starting in the lungs), where O_2 is finally consumed to produce adenosine triphosphate, the 'energy currency' molecule. As can be expected, mitochondrial content is typically augmented with endurance training.[33] In this respect, the capacity to extract energy from (i.e., to oxidize or 'burn') fatty acids and lactate in the mitochondria is also augmented in women's skeletal muscle fibers compared with those of men. Such an enhanced catabolic, specifically oxidative, capacity may be the consequence of increased mitochondrial content, given that no sex difference is noted when the comparison is adjusted by the density of mitochondria in skeletal muscle.[31]

Another recent piece of evidence indicates that mitochondrial oxidative capacity is similar in moderately endurance-trained women compared with highly endurance-trained men having a substantially higher (+33%) aerobic capacity.[32] In parallel to these cross-sectional data, acute endurance exercise (90 minutes of cycling at a controlled moderate exercise intensity relative to VO_{2max}) stimulates muscle mitochondrial biogenesis to a greater extent in women than men matched by the duration of endurance training per week.[34] Taken together, these findings denote that women may need lesser stimuli than men to increase mitochondrial content and function, thereby possessing higher potential to enhance their muscle metabolic oxidative capacity. As surprising as it may seem, women may indeed possess a superior combustion engine in their muscles, at least with respect to the capacity to catabolize key energy substrates. Stated differently, men are handicapped (relative to women) in the last step of the O_2 cascade.

WHY DO WOMEN HAVE SUPERIOR MUSCLE OXIDATIVE CAPACITY?

The use of fats to produce energy requires O_2. As we know, the capacity to deliver and consume O_2 (VO_{2max}) is reduced in women compared with men for a given physical activity level.[31] In other words, women and men subjected to the same training dose (i.e., duration multiplied by intensity relative to their maximal exercise capacity) achieve a different magnitude of improvement in VO_{2max}, what we named as the sex dimorphism in VO_{2max} trainability.[35] Possibly, the reduced central cardiac adaptations to endurance training and resultant blunted increases in O_2 delivery in women vs. men[36] may provide extra stimuli for the enhancement of skeletal muscle oxidative capacity in the former, leading to augmented mitochondrial content and function.

Parenthetically, higher mitochondrial content per muscle fiber does not necessarily translate into higher total number of mitochondria per unit of body (or leg) weight. Body composition consistently differs between sexes for a given endurance training status.[37] Women generally have a lower percentage of skeletal muscle in the trunk and limbs; therefore, the total number of mitochondria per total body weight might be similar or even be decreased compared with men. Provided that the percentage of muscle fibers that are activated during body-weight-bearing exercise is similar in both sexes, women would be subjected to a greater stress per muscle fiber: they have fewer muscle fibers for a given unit of body weight, which can be an additional stimulus contributing to enhance muscle oxidative capacity.

POTENTIAL EVOLUTIVE ROLE IN SEX DIFFERENCES IN MUSCLE OXIDATIVE CAPACITY

'Why' questions in biology generally require inferences about the course of evolution. In this respect, evolution has brought a prevalently lower body size in women relative to men. When the influence of body size or leg mass is statistically eliminated, sex differences in muscle mitochondrial content and oxidative capacity disappear.[31] Beyond humans, increased mitochondrial content and oxidative capacity has been found in skeletal muscle of animal species in which females have a smaller body size than males.[38, 39] Therefore, among mammalian species, the inverse relationship between body size and mitochondrial content in skeletal muscle appears to be the norm.[40]

The potential mechanistic link of body size and related anthropometrical variables with the skeletal muscle mitochondrial phenotype invites speculation. At first sight, the reduced overall dimensions of muscle fibers in women relative to men could contribute to differences in mitochondrial content and function, provided that the growth of muscle fibers and mitochondria is dissociated. In this regard, a peculiar characteristic of cellular biology may partly set apart the development of muscle fibers and mitochondria in women and men. Mitochondria are the only organelles in the human body that have their own genetic instructions, i.e., deoxyribonucleic acid (DNA), distinct from that which is present in the nucleus of most cells. In fact, mitochondrial DNA is in the vast majority of humans exclusively inherited from the mother, specifically from the maternal egg cell.[41] Mitochondrial genes are not subjected to substantial recombination; thus, whatever role these genes play in exercise performance, we cannot primarily blame to our father or to luck in DNA recombination. The maternal inheritance of mitochondria DNA may indeed be not attuned with the fast-evolutive inheritance of nuclear genes that determine the growth of skeletal muscle fibers in women and men (Figure 4.1). A mitochondrial 'genetic ceiling' might set the ultimate limit of muscle oxidative capacity, a ceiling closer to men than to women due to the greater muscle fiber size and contractile metabolic demands associated with the male sex.

Genetic differences may not be the only possible evolutive answer. The internal environment (partly determined by genes, epigenetics and external stimuli), including circulating proteins, hormones and metabolites, might stimulate further mitochondrial growth in women facilitating a type of metabolism more reliant on fat utilization during exercise,[42–44] as we will develop in the next chapter. Yet, the superior muscle oxidative capacity in women raises the obvious objection: why they generally display a reduced performance during prolonged endurance efforts compared with men.

DOES MUSCLE OXIDATIVE CAPACITY DETERMINE ENDURANCE PERFORMANCE?

From Chapter 1, it has been made clear that the rate-limiting step of VO_{2max} in healthy humans is O_2 delivery, not the final O_2 consumption in the mitochondria.[7, 45, 46] Given that VO_{2max} is a very strong determinant of exercise (endurance) capacity, the question naturally arises: do mitochondrial content and function matter? Well, endurance capacity, i.e., the potential to prolong exercise performance in humans, may be determined to a certain degree by multiple biological variables, in an independent or intertwined manner. Skeletal muscle mitochondrial content and function can be two of them. These variables, when enhanced, may provide a higher capacity to utilize abundant energy substrates (those that can be stored with little limitation in the human body, i.e., fats), thus theoretically providing advantages to sustaining aerobic metabolism during several hours of continuous moderate-intensity exercise. However, a straightforward answer to the aforementioned question cannot yet be provided, as mitochondria are not amenable to experimental manipulation in healthy humans conforming to current ethical standards. Relevant experiments must be performed in animals.

A series of experiments performed in rodents provides insight into this fundamental question, as far as the rat's physiology can be extrapolated to humans. In the pertinent study,[47] rats were made iron-deficient via a strict diet intervention, which diminished the activity of the iron-dependent enzymes that enable mitochondrial oxidation in their muscles. The activity of key mitochondrial oxidative enzymes in skeletal muscle was pruned down to 60–85% of normal levels. Likewise, skeletal muscle mitochondrial content was decreased by 30%. The iron-deficiency status also led to lower circulating hemoglobin, resulting in a 70% reduction in blood O_2 carrying capacity. As expected, VO_{2max} and endurance capacity, as determined via submaximal running to exhaustion in a treadmill, were largely decreased, the latter showing a 92% reduction compared with the control group (9 vs. 133 minutes). Subsequently, the iron-deficient rats were provided with a normal diet to replenish iron stores and correct the level of hemoglobin in blood and mitochondrial enzyme's activities. Blood hemoglobin was first corrected (at day 3 on the iron-rich diet) which paralleled the recovery of VO_{2max}. Mitochondrial oxidative capacity and endurance capacity were not fully recovered until day 7 on the iron-rich diet, which reinforces the dissociation between mitochondrial function and VO_{2max}, and between the latter and endurance capacity. Indeed, the progressive recovery of endurance capacity with the iron-rich diet followed closely that of mitochondrial oxidative capacity. Taken together, in rats, endurance capacity depends more on mitochondrial function than on VO_{2max}. Whether this statement can be extrapolated to humans remains an open question.

THE NEEDED EXPERIMENT

Notwithstanding structural and functional differences in skeletal muscle, the lower VO_{2max} in women—mainly determined by O_2 delivery factors previously considered (blood O_2 carrying capacity, blood volume, cardiac filling and output)[48–50]—may plausibly affect their sustained submaximal efforts. In fact, submaximal and maximal responses are tightly linked in the human body.[51] For a given submaximal O_2 consumption during moderate exercise, e.g., 35 mL/min/kg for fit endurance runners, those with substantially lower VO_{2max} (women) will experience greater homeostatic alterations. Namely, increases in circulating stress hormones, heart rate, central and peripheral neural activities affecting multiple organs will be augmented in women according to their reduced maximal aerobic exercise capacity. In short, the lower the VO_{2max}, the greater the stress of a given submaximal effort in absolute terms (i.e., not relative to VO_{2max}).

In order to reveal the importance of skeletal muscle oxidative capacity on sex differences in endurance performance, VO_{2max} or, more precisely, O_2 delivery to locomotive muscles should be matched between sexes. This must be done experimentally, as otherwise it will require higher training stimuli, hence possibly leading to increased skeletal muscle adaptations, in women relative to men.[35, 36] While this study remains to be performed, it does not seem outrageous to predict a superior endurance performance in women if their O_2 delivery to the locomotive muscles is the same as in men. Sex differences in skeletal muscle fatigability, which could be influenced by multiple intrinsic and/or neural-related factors, seems to depend primarily on O_2 availability.[52, 53] Once this variable is matched, the enhanced metabolic machinery in women's skeletal muscle fibers might emerge as a decisive factor leading to superior female endurance performance in prolonged (several hours) efforts.

REFERENCES

1. Young TR, Duncan BT, Cook SB. Evaluation of muscle thickness of the vastus lateralis by ultrasound imaging following blood flow restricted resistance exercise. *Clin Physiol Funct Imaging.* 2021;41(4):376–384.
2. Garg K, Corona BT, Walters TJ. Losartan administration reduces fibrosis but hinders functional recovery after volumetric muscle loss injury. *J Appl Physiol.* 2014;117(10):1120–1131.
3. Saltin B, Henriksson J, Nygaard E, Andersen P, Jansson E. Fiber types and metabolic potentials of skeletal muscles in sedentary man and endurance runners. *Ann N Y Acad Sci.* 1977;301:3–29.
4. Xu R, Andres-Mateos E, Mejias R, et al. Hibernating squirrel muscle activates the endurance exercise pathway despite prolonged immobilization. *Exp Neurol.* 2013;247:392–401.
5. Knight PK, Sinha AK, Rose RJ. *Effects of Training Intensity on Maximum Oxygen Uptake.* Davis, CA: ICEEP Publications; 1991.
6. Musch TI, Haidet GC, Ordway GA, Longhurst JC, Mitchell JH. Dynamic exercise training in foxhounds. I. Oxygen consumption and hemodynamic responses. *J Appl Physiol.* 1985;59(1):183–189.
7. Lundby C, Montero D, Joyner M. Biology of VO2 max: Looking under the physiology lamp. *Acta Physiol (Oxf).* 2017;220(2):218–228.
8. Fink WJ, Costill DL, Pollock ML. Submaximal and maximal working capacity of elite distance runners. Part II. Muscle fiber composition and enzyme activities. *Ann N Y Acad Sci.* 1977;301:323–327.
9. Morales-Alamo D, Losa-Reyna J, Torres-Peralta R, et al. What limits performance during whole-body incremental exercise to exhaustion in humans? *J Physiol.* 2015;593(20):4631–4648.
10. Morton RW, Sonne MW, Farias Zuniga A, et al. Muscle fibre activation is unaffected by load and repetition duration when resistance exercise is performed to task failure. *J Physiol.* 2019;597(17):4601–4613.
11. Lexell J, Henriksson-Larsen K, Sjostrom M. Distribution of different fibre types in human skeletal muscles. 2. A study of cross-sections of whole m. vastus lateralis. *Acta Physiol Scand.* 1983;117(1):115–122.

12. Serrano N, Colenso-Semple LM, Lazauskus KK, et al. Extraordinary fast-twitch fiber abundance in elite weightlifters. *PLOS ONE.* 2019;14(3):e0207975.
13. Gundersen K. Determination of muscle contractile properties: The importance of the nerve. *Acta Physiol Scand.* 1998;162(3):333–341.
14. Bottinelli R, Reggiani C. Human skeletal muscle fibres: Molecular and functional diversity. *Prog Biophys Mol Biol.* 2000;73(2–4):195–262.
15. Lexell J, Sjostrom M, Nordlund AS, Taylor CC. Growth and development of human muscle: A quantitative morphological study of whole vastus lateralis from childhood to adult age. *Muscle Nerve.* 1992;15(3):404–409.
16. Lexell J, Downham D, Sjostrom M. Distribution of different fibre types in human skeletal muscles. Fibre type arrangement in m. vastus lateralis from three groups of healthy men between 15 and 83 years. *J Neurol Sci.* 1986;72(2–3):211–222.
17. MacDougall JD, Sale DG, Alway SE, Sutton JR. Muscle fiber number in biceps brachii in bodybuilders and control subjects. *J Appl Physiol Respir Environ Exerc Physiol.* 1984;57(5):1399–1403.
18. Gonyea W, Ericson GC, Bonde-Petersen F. Skeletal muscle fiber splitting induced by weight-lifting exercise in cats. *Acta Physiol Scand.* 1977;99(1):105–109.
19. Wilson JM, Loenneke JP, Jo E, Wilson GJ, Zourdos MC, Kim JS. The effects of endurance, strength, and power training on muscle fiber type shifting. *J Strength Cond Res.* 2012;26(6):1724–1729.
20. Bathgate KE, Bagley JR, Jo E, et al. Muscle health and performance in monozygotic twins with 30 years of discordant exercise habits. *Eur J Appl Physiol.* 2018;118(10):2097–2110.
21. McNeil CJ, Vandervoort AA, Rice CL. Peripheral impairments cause a progressive age-related loss of strength and velocity-dependent power in the dorsiflexors. *J Appl Physiol.* 2007;102(5):1962–1968.
22. Graves JA. Evolution of vertebrate sex chromosomes and dosage compensation. *Nat Rev Genet.* 2016;17(1):33–46.
23. Nowak MA, Boerlijst MC, Cooke J, Smith JM. Evolution of genetic redundancy. *Nature.* 1997;388(6638):167–171.
24. Haizlip KM, Harrison BC, Leinwand LA. Sex-based differences in skeletal muscle kinetics and fiber-type composition. *Physiology (Bethesda).* 2015;30(1):30–39.
25. Alway SE, Grumbt WH, Gonyea WJ, Stray-Gundersen J. Contrasts in muscle and myofibers of elite male and female bodybuilders. *J Appl Physiol.* 1989;67(1):24–31.
26. Miller AE, MacDougall JD, Tarnopolsky MA, Sale DG. Gender differences in strength and muscle fiber characteristics. *Eur J Appl Physiol Occup Physiol.* 1993;66(3):254–262.

27. Blood donation: What to expect. https://www.mayoclinic.org/tests-procedures/blood-donation/about/pac-20385144. Accessed 13.12.2019, 2019.
28. https://www.sciencedaily.com/releases/2019/03/190327142058.htm, 2019.
29. Almog T, Almog O. *Academia: All the Lies: What Went Wrong in the University Model and What Will Come in its Place*. Independently Published; 2020.
30. Broad W, Wade N. *Betrayers of the Truth: Fraud and Deceit in the Halls of Science*. New York:Simon & Schuster; 1983.
31. Montero D, Madsen K, Meinild-Lundby AK, Edin F, Lundby C. Sexual dimorphism of substrate utilization: Differences in skeletal muscle mitochondrial volume density and function. *Exp Physiol*. 2018;103(6):851–859.
32. Cardinale DA, Larsen FJ, Schiffer TA, et al. Superior intrinsic mitochondrial respiration in women than in men. *Front Physiol*. 2018;9:1133.
33. Montero D, Cathomen A, Jacobs RA, et al. Haematological rather than skeletal muscle adaptations contribute to the increase in peak oxygen uptake induced by moderate endurance training. *J Physiol*. 2015;593(20):4677–4688.
34. Roepstorff C, Schjerling P, Vistisen B, et al. Regulation of oxidative enzyme activity and eukaryotic elongation factor 2 in human skeletal muscle: Influence of gender and exercise. *Acta Physiol Scand*. 2005;184(3):215–224.
35. Diaz-Canestro C, Montero D. Sex dimorphism of VO2max trainability: A systematic review and meta-analysis. *Sports Med*. 2019;49(12):1949–1956.
36. Diaz-Canestro C, Montero D. The impact of sex on left ventricular cardiac adaptations to endurance training: A systematic review and meta-analysis. *Sports Med*. 2020;50(8):1501–1513.
37. Ofenheimer A, Breyer-Kohansal R, Hartl S, et al. Reference values of body composition parameters and visceral adipose tissue (VAT) by DXA in adults aged 18–81 years-results from the LEAD cohort. *Eur J Clin Nutr*. 2020;74(8):1181–1191.
38. Colom B, Alcolea MP, Valle A, Oliver J, Roca P, Garcia-Palmer FJ. Skeletal muscle of female rats exhibit higher mitochondrial mass and oxidative-phosphorylative capacities compared to males. *Cell Physiol Biochem*. 2007;19(1–4):205–212.
39. Sandoval DA, Ryan KK, de Kloet AD, Woods SC, Seeley RJ. Female rats are relatively more sensitive to reduced lipid versus reduced carbohydrate availability. *Nutr Diabetes*. 2012;2:e27.

40. Hoppeler H. The different relationship of VO2max to muscle mitochondria in humans and quadrupedal animals. *Respir Physiol.* 1990;80(2–3):137–145.
41. Rius R, Cowley MJ, Riley L, Puttick C, Thorburn DR, Christodoulou J. Biparental inheritance of mitochondrial DNA in humans is not a common phenomenon. *Genet Med.* 2019;21(12):2823–2826.
42. Bennett E, Peters SAE, Woodward M. Sex differences in macronutrient intake and adherence to dietary recommendations: Findings from the UK Biobank. *BMJ Open.* 2018;8(4):e020017.
43. Lundsgaard AM, Kiens B. Gender differences in skeletal muscle substrate metabolism – molecular mechanisms and insulin sensitivity. *Front Endocrinol (Lausanne).* 2014;5:195.
44. Varlamov O, Bethea CL, Roberts CT, Jr. Sex-specific differences in lipid and glucose metabolism. *Front Endocrinol (Lausanne).* 2014;5:241.
45. Davies KJ, Maguire JJ, Brooks GA, Dallman PR, Packer L. Muscle mitochondrial bioenergetics, oxygen supply, and work capacity during dietary iron deficiency and repletion. *Am J Physiol.* 1982;242(6):E418–427.
46. Lundby C, Jacobs RA. Adaptations of skeletal muscle mitochondria to exercise training. *Exp Physiol.* 2016;101(1):17–22.
47. Davies KJ, Maquire JJ, Brooks GA, Dallman PR, Packer L. Muscle mitochondrial bioenergetics, oxygen supply, and work capacity during dietary iron deficiency and repletion. *Am J Physiol.* 1982;242(6):418–427.
48. Diaz Canestro C, Pentz B, Sehgal A, Montero D. Blood withdrawal acutely impairs cardiac filling, output and aerobic capacity in proportion to induced hypovolemia in middle-aged and older women. *Appl Physiol Nutr Metab.* 2021;[in print].
49. Diaz-Canestro C, Pentz B, Sehgal A, Montero D. Sex differences in cardiorespiratory fitness are explained by blood volume and oxygen carrying capacity. *Cardiovasc Res.* 2021;118(1):334–343.
50. Diaz-Canestro C, Siebenmann C, Montero D. Blood oxygen carrying capacity determines cardiorespiratory fitness in middle-age and older women and men. *Med Sci Sports Exerc.* 2021;53(11):2274–2282.
51. Pentz B, Diaz-Canestro C, Sehgal A, Montero D. Effects of blood withdrawal on cardiac, hemodynamic, and pulmonary responses to a moderate acute workload in healthy middle-aged and older females. *J Sci Med Sport.* 2021;25(3):198–203.
52. Ansdell P, Brownstein CG, Skarabot J, et al. Sex differences in fatigability and recovery relative to the intensity-duration relationship. *J Physiol.* 2019;597(23):5577–5595.
53. Hunter SK. Sex differences in fatigability of dynamic contractions. *Exp Physiol.* 2016;101(2):250–255.

Chapter 5

Fuel utilization and body composition

Abbreviations:

ATP, adenosine triphosphate
CO_2, carbon dioxide
H_2O, water
O_2, oxygen
VO_{2max}, maximal oxygen consumption

THE HUMAN METABOLIC ENGINE IN BRIEF

Energy extraction and utilization are determined by the attributes of the engine. As a case in point, let us briefly consider the process that takes place in a car's engine. Herein, the chemical energy present in the molecular bonds of the fuel is converted into heat (thermal energy) that is released in a fast fuel oxidation, i.e., combustion, which by means of the pistons is finally converted into motion (kinetic energy). The whole process is ignited by a spark plug (electrical energy) in a gasoline engine. In a diesel engine, the spark plug is not required, since the compression-induced heated air ignites the combustion. The remainder is pretty similar in gasoline and diesel engines. Yet, better not inject the wrong type of fuel in your car; the consequences would be disastrous. Apparently, trivial mechanical and chemical details make the difference between smooth energy processing or wreaking havoc in the engine.

In the human body, the oxidative process is lengthened and facilitated by the presence of proteins (enzymes) that function optimally at a specific temperature (~37°C). Multiple biochemical reactions transfer chemical energy between the broken and formed atomic bonds

DOI: 10.1201/9781003486893-5

until the end products of CO_2 and water (H_2O) are formed in the mitochondria in the presence of O_2. A parallel end product, ATP, stores part of the chemical energy in its bonds. The energy stored in ATP is used by all processes that require energy in our body, including muscular contraction and relaxation, ion transport across membranes and chemical (anabolic) reactions. The energy that is not ultimately transferred to ATP is gradually released in the form of thermal energy in exothermic reactions that partly determine body temperature. Little is wasted. In essence, the human metabolic engine turns chemical energy from external energy substrates into a 'gold chemical standard' (ATP), which represents the energy currency for biological work. Without ATP, and the biochemical pathways and primary energy substrates leading into it, any known form of life would not be conceivable.

ENERGY SUBSTRATES FOR HUMANS

The majority of existing chemical substances can be processed to some degree by the metabolic machinery of the human body. Regardless of their potential toxicity, energy may be extracted if the structure of these substances fit that of the proteins (enzymes) constituting our catabolic pathways. A conventional distinction is made between substances that positively contribute to biological processes, defined as nutrients, and those that do not. The blurry and controversial lines between positive, neutral and negative contributions to biological process necessarily leads to ambiguity. For instance, alcohol (ethanol) can be metabolized and energy extracted, 7 kcal per g. This is more than the energy extracted from nutrients like glucose or proteins (4 kcal per g).[1] Yet, alcohol is not considered a nutrient, since it interferes with multiple regulatory systems. On the other hand, insoluble dietary fiber, i.e., carbohydrate polymers that cannot be digested by human enzymes, belongs to the nutrient category. Despite providing no energy, dietary fiber in moderate amounts confer a myriad of benefits: normalization of bowel movement, attenuation of spikes in blood sugar by delaying the rate of glucose absorption into the circulation, decreased hunger associated with hypocaloric diets, among others. Ultimately, the categorization of a chemical substance as a nutrient or poison depends on physiological judgement, which always entails a degree of subjective interpretation.

Irrespective of nomenclatures, humans have learned what to eat—or at least what not to eat in order to avoid food poisoning and acute adverse effects on health—throughout 300,000 years of trial and error. At present, we are all born in a given culture with specific nutritional 'recommendations'. According to biochemical properties, the following broad groups of energy substrates, also known as macronutrients, are present in most diets: carbohydrates, fats and proteins. The latter play a relevant role only as energy substrate in very specific conditions: starvation, hypocaloric regimes, high-protein diets or very prolonged exercise. Under normal circumstances, carbohydrates and fats provide the bulk of energy. On average, 4 and 9 kcal are released by 1 g of carbohydrates and fats, respectively. Such is the amount of metabolizable energy, which is equal to the total energy in nutrients (determined by combustion) minus the energy lost in feces, urine, secretion and gases, a difference that defines nutrients' 'digestibility'. Hence, fats are energetically denser than carbohydrates. Yet, carbohydrates are essential for high-intensity exercise. The percentage contribution of carbohydrates to energy production exponentially increases with the intensity of exercise. Notably, at very high intensities (>90% of VO_{2max}), more than 80% of ATP must be obtained from the catabolism of carbohydrates. The aerobic catabolism of carbohydrate yields ATP, i.e., stored energy, at more than twice the pace of fats. In the absence of O_2, ATP can be produced at an even faster pace for short periods (~2 minutes) from the anaerobic catabolism of carbohydrates, in a process known as glycolysis. Hence, our metabolic engine requires carbohydrates to attain high rates of energy production.

STORAGE OF CARBOHYDRATES AND FATS

Given the importance of carbohydrates, the question of storage leaps to mind. How many carbohydrates can be stored in the body? Not many. In the archetypical healthy man with a body weight of 70 kg, approximately 500 g of carbohydrates can be stored in the form of glycogen (a polymer of glucose). The main tissue that serves as a reservoir for glycogen is skeletal muscle (400 g). The liver can also store a minor amount of glycogen (100 g). The chemical properties of glycogen strongly attract water: each g of glycogen links with 3 g of water—making glycogen a *bulky* energy substrate. The small amount of kcal available from the stored carbohydrates, i.e., 2000 kcal, contrasts with the vast amount available from fat reserves: 108,000 kcal derived from ~12 kg of metabolizable adipose tissue in our 70 kg man. Bearing in mind the energy that he would expend to run a marathon (~2700 kcal), fat could theoretically supply the energy for 40 marathons in a row, while not even one marathon could be completed only from the stored carbohydrates. Yet, the estimate of energy available from carbohydrates is 'inflated', since only the carbohydrate present in the exercising muscle fibers can be used for muscle's energy production. The reason is that once glucose enters into the muscle fiber, it cannot be released back into the circulation because muscle fibers lack the enzyme to revert the process. Hence, the glycogen stored in non-exercising muscles is inaccessible. Therefore, the human body mainly stores energy in the form of fat, specifically in the subcutaneous region and around internal organs. Carbohydrate ('high-octane') reserves are essentially residual, mainly available in the exercising muscles for relatively short and high-intensity efforts.

The limitation of carbohydrate storage would not be problematic for prolonged exercise performance if oral intake of carbohydrates could match the rate of consumption. This is not the case in humans. When glucose is directly administered in the circulation—hence bypassing the delay imposed by the gastrointestinal processing—the concentration of glucose in blood can be preserved and the time to fatigue delayed (~20 minutes) compared with oral glucose ingestion over a 3-hour, moderate-intensity exercise (between 70% and 80% of VO_{2max}).[2] As blood glucose concentration is kept constant via intravenous administration, 75% of glucose that is catabolized during exercise comes from the circulation, the remaining proceeding from intramuscular glycogen stores. Consequently, the more prominent limiting step in the delivery of glucose to the exercising muscles is the absorption of glucose in the gastrointestinal system. The glycogen stored in the liver and intramuscularly must provide the majority of glucose. Not surprisingly, endurance athletes are recommended to follow very high-carbohydrate diets, also including pre-competition meals and early snacks during exercise as well as afterwards (if competition follows the next day), all with high carbohydrate content. The nutritional goal is to magnify muscle glycogen reserves and delay as much as possible the decrease in blood glucose concentration and thereby carbohydrate availability to active muscles. When the 'high-octane' supply is diminished, so is performance, a decline that eventually must happen at high exercise intensities, in which active muscles consume more carbohydrates that they can assimilate, an imbalance ending in muscular fatigue and dysfunction.

SEX DIFFERENCES IN ENERGY SUBSTRATE UTILIZATION

In contrast with upstream steps in the O_2 transport and utilization chain, women do not apparently present any handicap in energy substrate utilization in skeletal muscle, as indicated in the previous chapter. Indeed, women exhibit augmented mitochondrial content in skeletal muscle compared with men matched by endurance performance and VO_{2max} (Figure 4.1).[3] This constitutional characteristic is associated with enhanced fatty acid and lactate oxidative capacity in women's skeletal muscle fibers.[3] Moreover, the overall (whole-body) energy substrate utilization during exercise intrinsically differs between women and men. Whole-body fat oxidation, as estimated from pulmonary analyses of O_2 consumption and CO_2 production, is higher, and carbohydrate oxidation correspondingly lower, in women compared with men matched by endurance training status and/or VO_{2max}.[4–11] The sex gap in whole-body energy substrate utilization is not trivial. Relative to men, women consume ~40% more fat per kg of body weight at a given exercise intensity (Figure 5.1).[3] Even at a high exercise intensity (85% of VO_{2max}), women largely rely on fat as energy substrate, lipid oxidation providing close to 40% of the energy production.[3] Of note, similar substrate utilization between sexes is observed prior to and after exercise, hence the remarkable sex difference in energy substrate utilization is confined to occur during exercise.[4] Women seem to turn into a 'diesel' engine at the time of physical exertion.

EXERCISE PERFORMANCE IMPLICATIONS

Prolonged endurance efforts, beyond the 3-hour barrier, may be strongly determined by factors other than VO_{2max}. Indeed, the most successful ever ultra-endurance runner, the Greek Yiannis Kouros, possessed a VO_{2max} of 'only' 62.5 mL/min/kg in his prime, i.e., one fourth less than the typical VO_{2max} in elite endurance athletes.[12, 13] Kouros did not consider himself the stronger athlete; indeed, he reported than his competitors were able to train harder than he did. What he was able to do apparently better than anyone was to consume calories (>13,000 kcal/day) during races lasting several days.[13] He achieved such an enormous energy consumption based

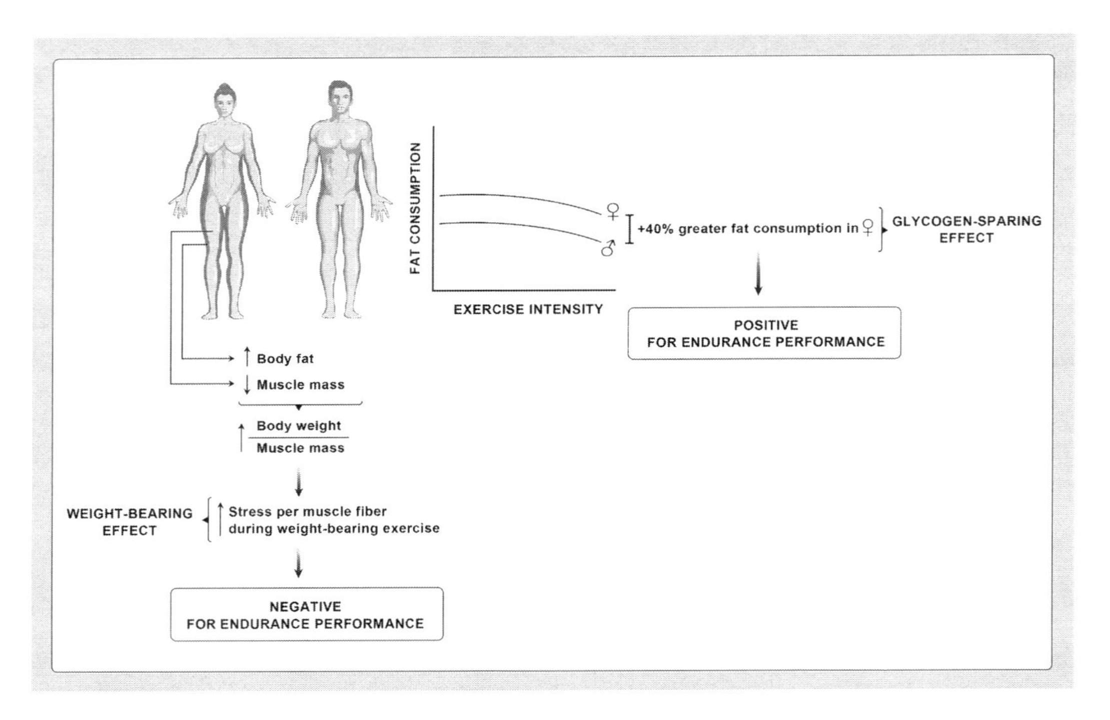

Figure 5.1 Sex differences in substrate utilization and body composition: influence on performance. The man and woman in the figure have the same body weight.

on his large macronutrient reservoirs complemented with high frequency snacks (every 15 minutes) almost exclusively comprised of carbohydrates (96%).[13] The gastrointestinal system of Kouros may have been able to digest and absorb carbohydrates into the circulation with ease, resulting in fewer gastrointestinal complications than those prevalently observed in ultra-endurance athletes,[14] thus facilitating a partial preservation of muscle glycogen. Nonetheless, the main macronutrient providing energy for ultra-endurance events is fat, given the low exercise intensities (<60% VO_{2max}) that can be maintained for hours and days.[13] As aforementioned, fat can be copiously stored in adipose tissue, which can be released into the circulation without the delay imposed by the gastrointestinal system—unlike carbohydrates, fat is a fuel that does not need to be ingested during exercise. The caloric consumption of Kouros during extremely prolonged exercise necessarily must also have entailed an outstanding metabolic capacity to burn his fat reservoirs.

As explained in the previous section, the capacity to use fat during exercise is augmented in the skeletal muscles of women. Compared with men, women can spare more of the limited intra- and extramuscular carbohydrate reserves at a given exercise intensity relative to their maximal exercise capacity. In theory, this metabolic characteristic should be beneficial for prolonged (several hours) efforts, in which carbohydrate reserves and delivery to skeletal muscle ultimately limit performance. In fact, evidence, albeit limited, indicates superior ultra-endurance (90 km) performance (average running speed) in women compared with men matched by their marathon's (42.2 km) performance.[15] In that study, women and men exhibited similar performance in shorter than marathon distances, at 10 and 21 km, where the metabolic 'ceiling' for carbohydrate reserves is not reached in most individuals.[15] It should be noted that at all distances, from 10 to 90 km, women exercised at a higher percentage of VO_{2max} (~60–85%) than men (~50–79%).[15] The enhance female capacity to use fats as energy substrate during high exercise intensities (up to 85% of VO_{2max}) plausibly facilitates that a higher fraction of women's aerobic capacity can be maintained for several hours of exertion.[3] Similar results are observed in swimming competitions in that women outperform men in the longest distances (~30 km),[16] albeit this could be explained by physical rather than physiological features such as (1) increased subcutaneous fat, thereby higher thermal insulation, and (2) enhanced buoyancy due to augmented body fat percentage and decreased density, leading to less resistance to move forward in the water, in women relative to men. Taken together, further experimental research should conclusively determine whether the enhanced fat catabolism of women during exercise confers crucial advantages in prolonged endurance efforts.

SEX-SPECIFIC BODY COMPOSITION: THE INGRAINED BURDEN OF WOMEN

In contrast to the structural advantages of body fat for long-distance swimming, the less body fat, the better for most land-based endurance efforts. Exercise on land typically involves a high vertical motion component, which is mainly opposed by gravity. Clear examples include long-distance running and uphill cycling, but in essence any prolonged effort that requires body motion on land (excluding downhill) is *assisted* by low body weight, which can be predominantly achieved by reducing body fat, while remaining muscularly fit. In fact, the interindividual variability in the metabolic cost of running at a fixed intensity (e.g., a concrete treadmill's speed and incline), as represented by O_2 consumption, is primarily explained (>90%) by body weight.[17] Even for cycling on a static cycle ergometer, i.e., an exercise with minimal vertical component, body weight determines more than 50% of the O_2 consumption at a given workload, plausibly attributed to the impact of lean body mass on basal O_2 consumption.[17] Not surprisingly given the lower metabolic capacity of fat compared with lean body mass (specifically muscle mass), body fat percentage is inversely associated with VO_{2max}, whose functionally relevant expression refers to O_2 consumption (mL/min) per kg of body weight.[18] The importance of body weight is certainly not overlooked by elite endurance athletes. In his autobiographic book *The Secret Race: Inside the Hidden World of the Tour de France*, the Olympic gold medalist and doping-convicted cyclist Tyler Hamilton stated that 'losing weight was the hardest but most effective way to increase the crucial watts by kilogram and, therefore, to succeed on the Tour de France'.[19]

Body composition is to some extent a sex-specific variable. Women have ~7–10 % more total body fat than men for a given overall fitness status.[20] The distribution of body fat also depends on sex.[21] The increased fat percentage in women compared with men is primarily attributed to higher subcutaneous fat in female limbs (arm, legs).[21] Relatively small sex differences in body fat are observed in the trunk.[21] Therefore, women accumulate more fat, and thereby more non-functional weight in the limbs (extremities), which according to physical principles must entail a higher energy cost for total body displacement in running-like activities (Figure 5.1).[22] Of note, the addition of as little as 100 g to runner's shoes increases the O_2 cost by up to 10% at fixed submaximal running velocities.[23] In essence, the higher the percentage of body weight distributed in distal parts of the body (subjected to greater angular movement), the higher the metabolic cost of a given displacement. In this regard, women have approximately 12% more fat in their legs compared with men matched by VO_{2max}.[21] Therefore, for an average two-leg weight of 20 kg, women carry 2400 g more than men in their extremities. It seems reasonable to argue that female endurance athletes would benefit most from the reduction of body fat and thereby weight. The question is to what extent. Whether women can decrease their body fat percentage to men's levels, without compromising exercise performance, remains to be scientifically ascertained. Among the athletes with the lowest body fat percentage, female bodybuilders (~10% body fat) do not to reach the lowest values of male counterparts (~3%), even assuming the likely use of 'body-sculping' drugs.[24] Human physiology may impose a stringent limit to the lowest body fat in healthy, or simply living, women. It is known that essential hormonal systems in the female's body are transitorily altered for several months in response to large reductions in body fat, possibly leading to chronic fatigue, anemia and metabolic disturbances.[25] While further evidence is needed, women's homeostasis seems to require a minimum percentage of body fat that is above the low percentage that can be sustained by male athletes.

FAT UTILIZATION AND BODY FAT: TWO SIDES OF THE SAME COIN

The excessive accumulation of body fat resulting in a body mass index > 30 kg/m^2 defines obesity. Substrate utilization is influenced by obesity. When obese and lean individuals are matched by age and VO_{2max} per lean body mass (a ratio accurately reflecting the dose of endurance training in obese and lean individuals), large differences in fat utilization during exercise are evident.[26] Obese individuals use 12% more fats, and proportionally fewer carbohydrates, at a moderate exercise intensity (50% of VO_{2max}) compared with lean individuals.[26] Specifically, obese individuals present with augmented oxidation of muscle triglycerides, therefore sparing muscle glycogen. Accordingly, obesity or high body fat percentage per se, in the absence of disease, entails improvements in substrate utilization during exercise. While this is speculative, it could be argued that the inherently higher body fat percentage in women underpins their increased capacity to use fat as a primary energy substrate. Conversely, a high body fat entails extra weight to be carried on, thus negatively affecting exercise performance. Hence, the metabolic advantage of women might come at a heavy price.

REFERENCES

1. Cederbaum AI. Alcohol metabolism. *Clin Liver Dis.* 2012;16(4):667–685.
2. Coggan AR, Coyle EF. Reversal of fatigue during prolonged exercise by carbohydrate infusion or ingestion. *J Appl Physiol.* 1987;63(6):2388–2395.
3. Montero D, Madsen K, Meinild-Lundby AK, Edin F, Lundby C. Sexual dimorphism of substrate utilization: Differences in skeletal muscle mitochondrial volume density and function. *Exp Physiol.* 2018;103(6):851–859.
4. Horton TJ, Pagliassotti MJ, Hobbs K, Hill JO. Fuel metabolism in men and women during and after long-duration exercise. *J Appl Physiol.* 1998;85(5):1823–1832.
5. Venables MC, Achten J, Jeukendrup AE. Determinants of fat oxidation during exercise in healthy men and women: A cross-sectional study. *J Appl Physiol.* 2005;98(1):160–167.
6. White LJ, Ferguson MA, McCoy SC, Kim H. Intramyocellular lipid changes in men and women during aerobic exercise: A (1) H-magnetic resonance spectroscopy study. *J Clin Endocrinol Metab.* 2003;88(12):5638–5643.
7. Tarnopolsky MA. Sex differences in exercise metabolism and the role of 17-beta estradiol. *Med Sci Sports Exerc.* 2008;40(4):648–654.
8. Blatchford FK, Knowlton RG, Schneider DA. Plasma FFA responses to prolonged walking in untrained men and women. *Eur J Appl Physiol Occup Physiol.* 1985;53(4):343–347.
9. Tarnopolsky LJ, MacDougall JD, Atkinson SA, Tarnopolsky MA, Sutton JR. Gender differences in substrate for endurance exercise. *J Appl Physiol.* 1990(68):302–308.
10. Phillips SM, Atkinson SA, Tarnopolsky MA, MacDougall JD. Gender differences in leucine kinetics and nitrogen balance in endurance athletes. *J Appl Physiol.* 1993;75(5):2134–2141.

11. Tarnopolsky MA, Atkinson SA, Phillips SM, MacDougall JD. Carbohydrate loading and metabolism during exercise in men and women. *J Appl Physiol.* 1995;78(4):1360–1368.
12. Lundby C, Robach P. Performance enhancement: What are the physiological limits? *Physiology (Bethesda).* 2015;30(4):282–292.
13. Rontoyannis GP, Skoulis T, Pavlou KN. Energy balance in ultramarathon running. *Am J Clin Nutr.* 1989;49(5 Suppl):976–979.
14. Arribalzaga S, Viribay A, Calleja-Gonzalez J, Fernandez-Lazaro D, Castaneda-Babarro A, Mielgo-Ayuso J. Relationship of carbohydrate intake during a single-stage one-day ultra-trail race with fatigue outcomes and gastrointestinal problems: A systematic review. *Int J Environ Res Public Health.* 2021;18(11):5737.
15. Speechly DP, Taylor SR, Rogers GG. Differences in ultra-endurance exercise in performance-matched male and female runners. *Med Sci Sports Exerc.* 1996;28(3):359–365.
16. Knechtle B, Dalamitros AA, Barbosa TM, Sousa CV, Rosemann T, Nikolaidis PT. Sex differences in swimming disciplines-can women outperform men in swimming? *Int J Environ Res Public Health.* 2020;17(10):3651.
17. Lundby C, Montero D, Gehrig S, et al. Physiological, biochemical, anthropometric, and biomechanical influences on exercise economy in humans. *Scand J Med Sci Sports.* 2017;27(12):1627–1637.
18. Vargas VZ, de Lira CAB, Vancini RL, Rayes ABR, Andrade MS. Fat mass is negatively associated with the physiological ability of tissue to consume oxygen. *Motriz The Journal of Physical Education.* 2018(4).
19. Hamilton T, Coyle D. *The Secret Race: Inside the Hidden World of the Tour de France.* New York: Bantam;2013.
20. Liguori D. et al. *ACSM's Guidelines for Exercise Testing and Prescription.* Philadelphia: Lippincott Williams & Wilkins; 2017.
21. Diaz-Canestro C, Pentz B, Sehgal A, Montero D. Sex dimorphism in cardiac and aerobic capacities: The influence of body composition. *Obesity (Silver Spring).* 2021;29(11):1749–1759.
22. Skinner HB, Barrack RL. Ankle weighting effect on gait in able-bodied adults. *Arch Phys Med Rehabil.* 1990;71(2):112–115.
23. Rodrigo-Carranza V, Gonzalez-Mohino F, Santos-Concejero J, Gonzalez-Rave JM. Influence of shoe mass on performance and running economy in trained runners. *Front Physiol.* 2020;11:573660.
24. van der Ploeg GE, Brooks AG, Withers RT, Dollman J, Leaney F, Chatterton BE. Body composition changes in female bodybuilders during preparation for competition. *Eur J Clin Nutr.* 2001;55(4):268–277.
25. Hulmi JJ, Isola V, Suonpaa M, et al. The effects of intensive weight reduction on body composition and serum hormones in female fitness competitors. *Front Physiol.* 2016;7:689.
26. Goodpaster BH, Wolfe RR, Kelley DE. Effects of obesity on substrate utilization during exercise. *Obes Res.* 2002;10(7):575–584.

Chapter 6

Training adaptations

Abbreviations:

ET, endurance training
Hb, hemoglobin
LBM, lean body mass
LV, left ventricle
O_2, oxygen
Q_{max}, maximal cardiac output
TPR, total peripheral resistance
VO_{2max}, maximal oxygen consumption

HUMAN TRAINABILITY

Our species demonstrates a wide spectrum of VO_{2max}, from ~20 to 85 $mL \cdot min^{-1} \cdot kg^{-1}$ in healthy adults.[1–3] A fraction of this variability can be explained by different levels of physical activity and/or genetic predisposition. In fact, differences in VO_{2max} among Olympic champions in endurance events, presumably with similar (optimal) levels of training and genetically predisposed to excel in endurance, are limited to a very narrow range (±2.5 $mL \cdot min^{-1} \cdot kg^{-1}$).[1] On the other hand, training studies in the general population have consistently demonstrated that a certain dose of endurance training (ET) does not induce uniform gains in VO_{2max} at the individual level, one third of endurance training ET-induced changes in VO_{2max} being scattered more than one standard deviation from the mean.[4–7] Notably, some healthy individuals seemingly do not respond to ET, thus they are labeled as 'non-responders'. The prevalence of VO_{2max} non-responders has been claimed to reach up to 20% in healthy individuals—i.e., one out of five would never

DOI: 10.1201/9781003486893-6

improve VO_{2max} with ET, presumedly due to genetic determinism.[8] Nonetheless, caution should be given to such a stunning claim, particularly when coming from academic circles not necessarily receptive and externally checked according to facts.

Human trainability is not atemporal; it varies over time in the same individual. Indeed, overtrained individuals are not expected to respond to training. Besides biological- and lifestyle-related variability, any measurement entails a typical error. VO_{2max} is no exception, with a typical error of measurement of ~5%.[9] Consequently, at the individual level we can never be highly confident in that a moderate (≤5%) change in VO_{2max} is a 'true' adaptation or the result of methodological error. Nonetheless, we may enhance the reliability via duplicate-intervention measurements. This approach is commonly recommended in the clinical arena for the testing of drugs, in which within-subject response variability is a well-known source of confounding.[10, 11]

Carsten Lundby, a Danish physiologist, applied duplicate-intervention measurements for the assessment of VO_{2max} trainability.[12] Specifically, he sought to determine whether the VO_{2max} response to ET was essentially dose-dependent. To this end, healthy, moderately active young men were divided into five training groups respectively comprising 1, 2, 3, 4 and 5 × 60-minute ET sessions per week but otherwise following an identical ET program for 6 weeks—an initial period of time in that most of the potential VO_{2max} improvement in the medium term (up to ~2 years) is elicited.[13–16] Non-response was primarily defined as any change in maximal incremental cycling power output (W_{max}, a robust marker of exercise capacity strongly correlated with VO_{2max}) within the typical error of measurement (±3.96%); nonetheless, similar results were found using W_{max} or VO_{2max} as the main outcome of the response to ET. Non-responders to the first 6-week ET program completed a duplicate 6-week ET program including two additional exercise sessions per week. Following the first 6-week ET block, the prevalence of VO_{2max} 'non-response' was 69%, 40% and 29% in individuals exercising 60, 120 or 180 minutes per week, respectively, and was absent (0%, i.e., all individuals responded to training) in those exercising for 240 or 300 minutes per week. Moreover, all non-responders did improve VO_{2max} in the second 6-week ET block with increasing training dose (+120 minutes per week). Hence, the dose-dependent nature of VO_{2max} responses to ET was demonstrated between and within healthy individuals. Notwithstanding, a degree of variability higher than the typical error of measurement was present between individuals for a given dose of ET, which still denotes the presence of considerable inter-individual variability in VO_{2max} trainability.

SEX DIFFERENCES IN TRAINABILITY: POOLED EVIDENCE ON AEROBIC AND CARDIAC ADAPTATIONS

Among the individuals exposed to the highest ET dose, i.e., endurance athletes, women commonly present with ~10 $mL \cdot min^{-1} \cdot kg^{-1}$ less VO_{2max} than men.[1, 17] However, whether the ET dose is similar between female and male endurance athletes is uncertain, albeit remarkably, a similar sex gap in aerobic capacity is also present among sedentary individuals.[18] In addition to cross-sectional evidence, which can be largely affected by potential confounding factors, a meta-analysis of ET studies did find sex differences in VO_{2max} responses.[19] After pooling the effects of 8 ET interventions (ranging from 7 to 52 weeks of duration) each comprised by healthy men and women matched by age and physical activity status (untrained), VO_{2max} improvements were greater (+2 mL/min/kg, + 40 %) in men relative to women.[19] The total time and average relative intensity of ET stimuli, i.e., the dose of ET, did not differ between sexes. Moreover, the higher VO_{2max} response in men relative to women was independent of the duration of ET, some of the studies involving long-term (1 year) interventions in which most of the potential medium-term improvement in VO_{2max} is certainly reached. These findings strongly imply the presence of a sexual dimorphism in the physiology of VO_{2max} improvement.

Cardiac structure and function must be first looked upon when looking for potential major mechanisms underlying differences in VO_{2max}. In a recent meta-analysis, the influence of sex on cardiac adaptations to ET was investigated.[20] Specifically, the focus was placed on the main cardiac pumping chamber, the left ventricle (LV), whose structural and functional adaptations strongly correlate with VO_{2max}.[21] Data from 26 exercise training studies comprising a similar dose of ET in healthy untrained and age-matched women and men were pooled. The duration of ET programs ranged from 12 weeks (a prudent minimal period of time for LV adaptations to be present) to 1 year. The internal volume of the LV was augmented to a greater extent in men compared with women (+16 vs. +5 mL), while the amount of blood pumped per heart beat, i.e., the stroke volume, was only significantly increased in men (+9 mL). Yet, the mass of the LV was similarly enhanced (+10 vs. +7 g) in men and women. Hence, women, relative to men, demonstrate a greater increase in LV mass than internal volume, resulting in LV concentric hypertrophy, following medium- to long-term ET. As previously developed in Chapter 2, such type of cardiac remodeling goes in the opposite direction from the one leading to increased maximal cardiac output (Q_{max}) and VO_{2max}.[21] An eccentrically enlarged LV with normal thickness and supple walls is the preferred phenotype to achieve the highest cardiac function, as represented by Q_{max}. While sex differences in trainability may be present beyond the heart, delving into and elucidating what constrains favorable cardiac adaptations in women is a fundamental step in the attempt to ascertain why their improvement in aerobic capacity is blunted.

WHY OPTIMAL CARDIAC ADAPTATIONS TO ENDURANCE TRAINING ARE LIMITED IN WOMEN?

Provided that ET delivers a similar dose of exercise in both sexes, biological factors must explain the limited cardiac adaptability in women. The external fibrous and quasi-non-distensible layer of the heart, i.e., the pericardium, can be one of these factors. The pericardium determines the upper (structural) limit of overall cardiac volume. Given that the pericardium area is smaller in women compared with men, absolute cardiac expansion can be expected to be constrained in the former. Nonetheless, the pericardium should also restrain eccentric myocardial adaptations in men, yet such a limitation is substantially less pronounced.[20] Could there be sex differences in the distensibility of the pericardial tissue? To date, no conclusive evidence has been reported in this regard.

External to the pericardium, the female heart may also encounter limitations for cardiac expansion. The majority of space in the rib cage is occupied by the lungs. The available 'room' for cardiac expansion thus partly depends on lung size and function. In this regard, women require higher respiratory work compared with men for a given absolute ventilatory flow during moderate- and high-intensity exercise, due to smaller female airways (see Chapter 1).[22, 23] Accordingly, higher intrathoracic pressures must be produced by women's expiratory muscles in order to breathe out during exercise, which negatively impact on cardiac filling and thereby cardiac size. Furthermore, cardiac wall tension may reasonably be elevated in exercising women relative to men owing to higher pressures in the rib cage. Limited cardiac expansion and increased wall stress favors cardiac concentric, rather than eccentric, adaptations to exercise in women.[24, 25]

Cardiovascular structure and the regulation of blood volume are reciprocally related. In fact, the maximal anatomical size of the cardiovascular system is the ultimate limiting factor of blood volume. Besides being a physical barrier, cardiovascular structure also contributes to the endocrine regulation of blood volume. The presence of potent blood volume-regulating hormones released by the cardiac atria in response to changes in cardiac filling was first reported by Argentinian and Canadian cardiovascular researchers in the 1980s—a relatively recent discovery in the history of medicine.[26] Adolfo Jose de Bold and colleagues discovered that in response to increased cardiac filling, as sensed by augmented atrial volume, myocardial cells in the atria release natriuretic peptides to the circulation. The main effect of natriuretic peptides is the reduction of blood volume via signaling the kidneys to increase sodium (and thereby water) excretion. Reduced blood volume concomitantly decreases cardiac filling and hence atrial volume, which diminishes the stimuli to release natriuretic peptides to the circulation, in a feedback loop that stabilizes atrial and overall cardiac volume. Importantly, the stimuli sensed by the myocardial cells in the atria partly depends on cardiac structure and function; i.e., the smaller the atrial volume in basal conditions, the larger the stimuli induced by given absolute increase in cardiac filling. Despite the fact that endocrine systems are generally adjusted according to organ size, the circulating level of natriuretic peptides is elevated in women compared with men in the general population, which may be partly explained by the smaller female heart.[27] The cardiac phenotype may inherently contribute to the regulation of blood volume, namely to its down-regulation in women to a greater extent than in men. Following this rationale, cardiac adaptations to ET in women would promptly hit the 'wall' that constrain adaptations in the fundamental determinant of cardiac pumping capacity, i.e., the filling of the heart.

EARLY CHILDHOOD TRAINING: A POTENTIAL WINDOW OF ENHANCED CARDIAC ADAPTABILITY

In multiple aspects, living beings are more malleable the younger they are. The heart seems to be no exception. A study in rats tested the hypothesis that early ET (during the 'juvenile' period in rats) leads to chronic modification of cardiac structure.[28] Rats were grouped according to their age (5 weeks for 'juvenile', 11 weeks for 'adolescence', 20 weeks for 'adulthood') and subjected to an identical 4-week ET dose (1 hour of running at a speed of 20 m/min, 5 days/week) on a specifically designed treadmill—no differences in body weight or total physical activity were initially noted between groups, suggesting that physical development was almost completed in rats assigned to the youngest ('juvenile') group. Following the training intervention, rats in 'juvenile' and 'adolescence' groups remained sedentary until they reached middle age (24 weeks), at which point they were sacrificed along with rats in the 'adulthood' group in order to examine cardiac structure. In addition, a subgroup of rats in the 'juvenile' group were sacrificed right after the training intervention (9 weeks) and compared with sedentary rats of the same age. It should be noted that a 4-week period of training in rats equates to ~2 years of training in humans, given the relatively short lifespan (3 years) and accelerated metabolism of rats relative to humans. From such a time perspective, the result of increased LV mass and total number of myocardial fibers (cardiac muscle fibers) after 'juvenile' training compared with sedentary lifestyle is not surprising. What was unexpected occurred at middle age (24 weeks). Middle-aged rats that were trained during the 'juvenile' period (and afterwards remained sedentary) preserved higher LV mass and myocardial fibers compared with lifelong sedentary rats. Moreover, LV responses to training were progressively diminished with older age. In fact, 'adulthood' training resulted in attenuated gains in LV mass and no increase in myocardial fibers relative to lifelong sedentary rats. Likewise, the increase in myocardial fibers was of lesser magnitude with 'adolescence' than with 'juvenile' training. Taken together, early ('juvenile') ET maximize adaptations in cardiac structure that are not achievable, or might require extra stimuli, at a later age in rats.

Whether these findings can be extrapolated to humans, and particularly to women, has yet to be further investigated, considering the quadrupedal-specific nature of the cardiovascular system in rats and the scarce evidence available.[29, 30] Long-term (≥2 years) studies including female and male children/adolescents allocated to ET and control groups, along with parallel ET-dose matched groups in adults, are needed. In humans, however, the aerobic exercise capacity is generally higher in adults than in children and in men versus women; thus the ET dose (total duration × average intensity) should be matched between groups by the product of duration and intensity-relative terms, for instance, relative to VO_{2max}. This may not be a major issue, provided that training stimuli are determined by relative rather than absolute doses.[12] A question that could be additionally answered using the aforementioned study design is whether the limited responsiveness to ET stimuli of the female versus male heart depends on age. If no sex difference in cardiac adaptations is observed in children/adolescents but lower female cardiac adaptability is confirmed in adulthood, it may be that at some period before adulthood, women are less physically active than men,[31] thus explaining lesser profitability of the 'enhanced cardiac adaptability window' to increase the number of myocardial fibers, which may partly determine the structural and functional potential of the heart.[28] Sadly, this might entail lower cardiac capacity and training response for the remained adult lifetime in women.

INCREASING BLOOD O_2 CARRYING CAPACITY: THE ESCAPE ROUTE FOR WOMEN?

Along with Q_{max}, the capacity to carry O_2 in blood is a major determinant of aerobic exercise capacity. In fact, when hemoglobin (Hb) concentration (i.e., the concentration that determines how much O_2 can be carried in blood) is altered by a given percentage, a corresponding percentage change in VO_{2max} is observed.[32] In this respect, experimental studies altering blood O_2 carrying capacity have demonstrated a similar impact on VO_{2max} in women and men.[32] As previously developed in Chapter 3, women consistently present a ~10% lower Hb concentration than men, irrespective of age and cardiorespiratory fitness status.[32, 33] Hence, in theory women have a greater range to increase blood O_2 carrying capacity and thereby VO_{2max}. In healthy women, Hb concentration can be largely augmented with pharmacological interventions, e.g., via erythropoietin administration.[34] Yet, neither ET nor any known type of exercise intervention increases Hb concentration in women or men.[13, 35] It should be noted that total circulating Hb is generally enhanced by ET in an approximate or lower proportion to plasma volume expansion, which keeps Hb concentration unaltered or slightly reduced.[12, 13, 16] Normal physiology thus limits increases in blood O_2 carrying capacity beyond a sex-specific range.[35]

There is at least one exception to the previous assertion. Altitude training, referring to living and (possibly but not necessarily) training at high altitude relative to sea level is known to increase blood O_2 carrying capacity.[36–38] Specifically, Hb concentration is elevated within hours of exposure to natural or simulated high altitude (>2500 m).[38] Such a very short-term effect is exclusively due to plasma volume reduction, since red blood cell formation (i.e., erythropoiesis) is a slow process that usually requires at least 2–3 weeks to induce detectable increases in red blood cell volume.[38] Therefore, blood volume is reduced during the first days of high-altitude exposure, which should entail lower cardiac filling and thereby impaired Q_{max} due to the law of the heart, as explained in Chapter 2. In fact, endurance athletes adding altitude training to their ET program may incorporate living and ET sojourns (2–6 weeks) at high altitude a few weeks before a major competition. The impact of such high altitude stages on blood O_2 carrying capacity is on average moderate (~3% to 10% increment in Hb concentration) and highly variable between studies and individuals.[39, 40] The contribution of erythropoiesis to high altitude-induced increment in Hb concentration in the medium-term may be similar or larger than that attributed to plasma volume reduction, provided the following expected adaptations: (1) a ~100 mL increase in red blood cell volume per every 2 weeks of high altitude exposure, and (2) unaltered total blood volume following several (≥3) weeks at high altitude.[38, 39, 41] The altitude effect on Hb concentration increases with the altitude level.[38] However, the absolute ET stimuli is concomitantly decreased in proportion to the reduction of O_2 availability at high altitude, leading to partial loss (de-training) of other ET adaptations (e.g., muscular-related adaptations).[42] A compromise is therefore made by endurance athletes exposed to high altitude, which, in order to gain an hematological edge (the key factor for their success), must somewhat 'descend' from peak form in other systems, organs and tissues.

The increase in red blood cells induced by weeks of exposure to high altitude is short-lasting.[43] Erythropoiesis is immediately slowed down when the hypoxic stimuli is attenuated or abolished with the return to lower altitude, leading to the re-establishment of normal blood O_2 carrying capacity in ~10 days after the high-altitude sojourn.[43] Therefore, unless the high-altitude-induced gain in red blood cells is withdrawn, preserved and eventually reinfused (i.e., blood doping), our physiology allows only for a fleeting improvement in blood O_2 carrying capacity. Such a time constraint, requiring high-altitude training camps to be planned very close to competitions, along with the concomitant hypoxic-related reduction of absolute training dose, greatly limits the interest of adding high altitude for performance enhancement.[44]

As it happens with most topics in human physiology, studies on altitude training have mainly included men.[44] In a recent systematic review on the effects of altitude training on hematological variables, approximately one third of selected studies included a meaningful sample of women and no relevant sex comparison was reported.[45] Consequently, whether (or the extent to which) the aforementioned discussion applies to women remains uncertain. Nonetheless, sex differences in the hematological response to high altitude might be hypothesized according to clinical evidence of the sex-specific impact of prolonged high altitude exposure on the 'excessive erythropoiesis' condition, with women being largely less affected than men possibly due to the effects of estrogen.[46] Future studies are needed to elucidate the potential of altitude-related strategies to improve aerobic capacity in women.

TARGETING THE PERIPHERY TO ENHANCE WOMEN'S CARDIAC CAPACITY

The composition and function of tissues in the limbs (legs, arms) are not considered to determine aerobic capacity in healthy individuals. As a case in point, the amount of skeletal muscle in the exercising limbs does not limit VO_{2max}, according to studies in men.[47, 48] Once a large fraction (~50%) of total muscle mass is activated during incremental exercise (e.g., with leg cycling exercise), whole-body Q_{max} and VO_{2max} are mainly attained.[47, 48] Yet, women and men largely differ in total muscle mass in relative and absolute terms. Compared with women, men matched by age and physical activity possess around ~8–10% more total lean body mass (LBM)—a readily measurable variable of body composition for the most part comprised of skeletal muscle mass—and the percentage difference is almost doubled in the limbs.[49] In absolute terms, a ~15–20 kg sex gap in total LBM is found on average between sexes.[49] Likewise, little sex overlap is observed in adults of European American ethnicity, with women and men predominantly falling in the lower (<50 kg total LBM) and upper (>50 kg total LBM) segments of the entire LBM range, respectively. Such a large quantitative gap in LBM may entail qualitative differences in the sex-specific limiting role of LBM for cardiac and aerobic capacities, as developed hereunder.

The cardiac capacity to pump blood to the systemic circulation partly depends on the resistance encountered by the left ventricle to eject blood.[50–52] Such a resistance, known as total peripheral resistance (TPR), is reduced in proportion to the overall degree of arterial vasodilation in the systemic circulation. In this respect, skeletal muscle plays a prominent role. During exercise, the greatest fraction of cardiac output (Q) (85–90%) flows towards active skeletal muscle, which is greatly facilitated by the vasodilation of blood vessels irrigating active muscles. The smallest arteries (arterioles) in skeletal muscle have the largest capacity to modify its vascular tone in response to exercise-induced neurohumoral stimulation and muscle contraction-derived metabolites, which are released into the circulation in proportion to muscle catabolism. TPR during exercise is thus mainly determined by the number and degree of vasodilation of arterioles in active skeletal muscle. Provided that women and men do not differ in the number and function of arterioles per unit of skeletal muscle mass,[53] the lesser the quantity of total muscle mass, the lower the theoretical capacity to vasodilate and thereby decrease TPR. In fact, TPR at rest and during exercise is elevated in women compared with men throughout the adult lifespan.[54, 55] The lower skeletal muscle in women may thus impose a limitation to the potential reduction in TPR, thus constraining Q_{max} and VO_{2max} (Figure 6.1).

Recent investigations have indeed found strong relationships between LBM, TPR, Q_{max} and VO_{2max} in women but not in age- and physical activity-matched men.[49] As to why these associations are not present in men, the amount of LBM in men provides a plausible answer. Indeed, no association is generally expected between LBM and TPR in men, since the vast majority possess a variable *excess* of muscle mass.[56] The limiting role of LBM might be more prevalently revealed in those with lower skeletal muscle, i.e., women. Perhaps the lower amount of skeletal muscle in women may entail maximal (or near maximal) fiber recruitment and TPR reduction in women's

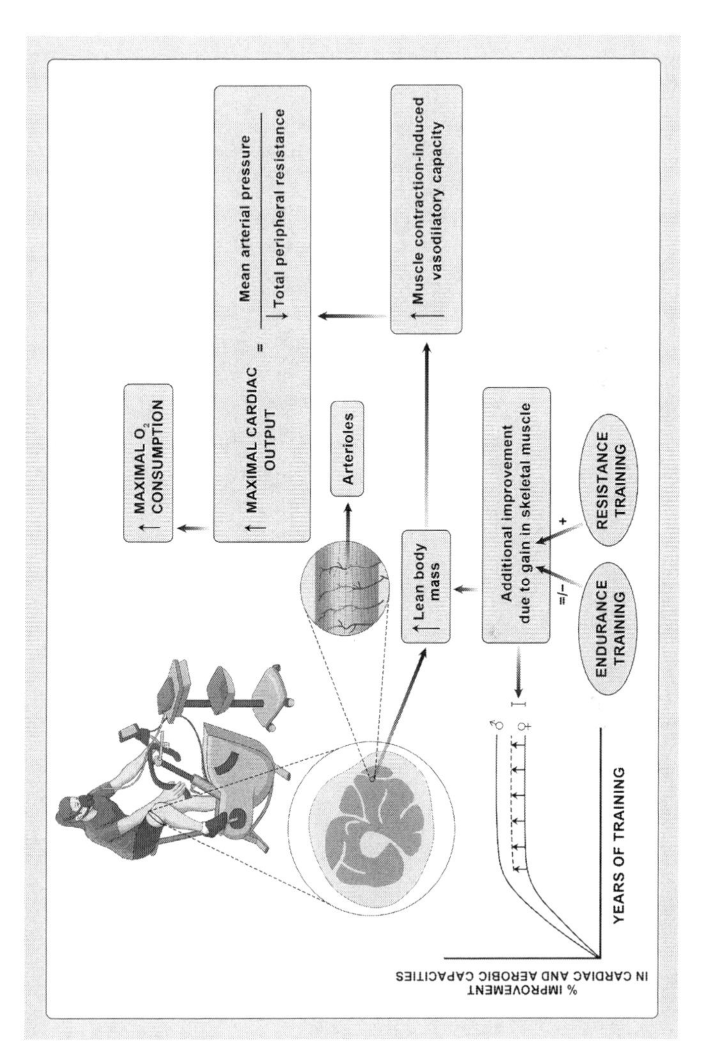

Figure 6.1 Potential role of skeletal muscle gain on cardiac and aerobic capacities in women.

active muscles during incremental exercise, which may not be the case for men. Presumably, if men were able to vasodilate all their arterioles in active skeletal muscles during leg cycling or whole-body exercise, normal blood pressure could not be maintained, and fainting would plausibly occur.[57] It should be noted that the aforementioned investigations included European American individuals, the major ethnic group with the highest LBM. In this regard, in East Asian (Hans Chinese) individuals, in which low LBM is the norm, LBM is associated with TPR, Q_{max} and VO_{2max} in both women and men.[58] Low LBM, rather than sex, seems to be the primary causative factor.

Following the aforementioned rationale, the question arises: would the cardiac and aerobic capacities of women (and men with low LBM) be enhanced with increasing LBM, specifically involving the accretion of skeletal muscle mass (Figure 6.1)? Thus far, the question remains mainly unanswered. Scarce evidence from a resistance training study performed in the late 1970s indicates substantial increments (+11%) in VO_{2max} in women, but not men, despite both women and men exhibited similar gains in LBM.[59] Nonetheless, provided that this finding is confirmed in further studies, increasing LBM may have only a moderate impact on cardiac capacity, since the 'enclosed' heart is more limited in its potential enlargement than skeletal muscles. Consequently, high LBM, if not accompanied by low body weight (thinness), cannot maximize VO_{2max} and endurance performance—as stereotypically exemplified by bodybuilders. That notwithstanding, a recent observation reveals that women with the highest VO_{2max}—i.e., female elite endurance athletes—are generally 'heavier' (greater body mass index) than their male counterparts, a gap that may not be entirely explained by sex differences in body fat percentage.[60, 61] Women may thus be naturally enforced to be 'bulkier' than men relative to body height for a given aerobic exercise capacity.

REFERENCES

1. Lundby C, Robach P. Performance enhancement: What are the physiological limits? *Physiology (Bethesda)*. 2015;30(4):282–292.
2. Montero D, Diaz-Canestro C. Endurance training and maximal oxygen consumption with ageing: Role of maximal cardiac output and oxygen extraction. *Eur J Prev Cardiol*. 2016;23(7):733–743.
3. Montero D, Diaz-Canestro C, Lundby C. Endurance training and VO2max: Role of maximal cardiac output and oxygen extraction. *Med Sci Sports Exerc*. 2015;47(10):2024–2033.

4. Lortie G, Simoneau JA, Hamel P, Boulay MR, Landry F, Bouchard C. Responses of maximal aerobic power and capacity to aerobic training. *Int J Sports Med.* 1984;5(5):232–236.
5. Bacon AP, Carter RE, Ogle EA, Joyner MJ. VO2max trainability and high intensity interval training in humans: A meta-analysis. *PLOS ONE.* 2013;8(9):e73182.
6. Bouchard C, Sarzynski MA, Rice TK, et al. Genomic predictors of the maximal O(2) uptake response to standardized exercise training programs. *J Appl Physiol.* 2011;110(5):1160–1170.
7. Cohen L, Holliday M. *Statistics for Education and Physical Education. Inferential Statistics.* London: Harper; 1979.
8. Timmons JA, Knudsen S, Rankinen T, et al. Using molecular classification to predict gains in maximal aerobic capacity following endurance exercise training in humans. *J Appl Physiol.* 2010;108(6):1487–1496.
9. Shephard RJ, Rankinen T, Bouchard C. Test-retest errors and the apparent heterogeneity of training response. *Eur J Appl Physiol.* 2004;91(2–3):199–203.
10. Hecksteden A, Kraushaar J, Scharhag-Rosenberger F, Theisen D, Senn S, Meyer T. Individual response to exercise training - a statistical perspective. *J Appl Physiol.* 2015;118(12):1450–1459.
11. Senn S, Rolfe K, Julious SA. Investigating variability in patient response to treatment--a case study from a replicate cross-over study. *Stat Methods Med Res.* 2011;20(6):657–666.
12. Montero D, Lundby C. Refuting the myth of non-response to exercise training: 'non-responders' do respond to higher dose of training. *J Physiol.* 2017;595(11):3377–3387.
13. Montero D, Breenfeldt-Andersen A, Oberholzer L, et al. Erythropoiesis with endurance training: Dynamics and mechanisms. *Am J Physiol Regul Integr Comp Physiol.* 2017;312(6):R894–R902.
14. Montero D, Cathomen A, Jacobs RA, et al. Haematological rather than skeletal muscle adaptations contribute to the increase in peak oxygen uptake induced by moderate endurance training. *J Physiol.* 2015;593(20):4677–4688.
15. Montero D, Lundby C. Regulation of red blood cell volume with exercise training. *Compr Physiol.* 2018;9(1):149–164.
16. Lundby C, Montero D, Joyner M. Biology of VO2 max: Looking under the physiology lamp. *Acta Physiol (Oxf).* 2017;220(2):218–228.
17. Rusko HK. Development of aerobic power in relation to age and training in cross-country skiers. *Med Sci Sports Exerc.* 1992;24(9):1040–1047.
18. Bouchard C, Daw EW, Rice T, et al. Familial resemblance for VO2max in the sedentary state: The HERITAGE family study. *Med Sci Sports Exerc.* 1998;30(2):252–258.
19. Diaz-Canestro C, Montero D. Sex dimorphism of VO2max trainability: A systematic review and meta-analysis. *Sports Med.* 2019;49(12):1949–1956.

20. Diaz-Canestro C, Montero D. The impact of sex on left ventricular cardiac adaptations to endurance training: A systematic review and meta-analysis. *Sports Med.* 2020;50(8):1501–1513.
21. Levine BD. VO2max: What do we know, and what do we still need to know? *J Physiol.* 2008;586(1):25–34.
22. Dominelli PB, Molgat-Seon Y, Bingham D, et al. Dysanapsis and the resistive work of breathing during exercise in healthy men and women. *J Appl Physiol.* 2015;119(10):1105–1113.
23. Molgat-Seon Y, Dominelli PB, Ramsook AH, et al. The effects of age and sex on mechanical ventilatory constraint and dyspnea during exercise in healthy humans. *J Appl Physiol.* 2018;124(4):1092–1106.
24. Diaz-Canestro C, Montero D. Female sex-specific curtailment of left ventricular volume and mass in HFpEF patients with high end-diastolic filling pressure. *J Hum Hypertens.* 2020.
25. Regitz-Zagrosek V, Kararigas G. Mechanistic pathways of sex differences in cardiovascular disease. *Physiol Rev.* 2017;97(1):1–37.
26. de Bold AJ, Borenstein HB, Veress AT, Sonnenberg H. A rapid and potent natriuretic response to intravenous injection of atrial myocardial extract in rats. *Life Sci.* 1981;28(1):89–94.
27. Suthahar N, Meijers WC, Ho JE, et al. Sex-specific associations of obesity and N-terminal pro-B-type natriuretic peptide levels in the general population. *Eur J Heart Fail.* 2018;20(8):1205–1214.
28. Asif Y, Wlodek ME, Black MJ, Russell AP, Soeding PF, Wadley GD. Sustained cardiac programming by short-term juvenile exercise training in male rats. *J Physiol.* 2018;596(2):163–180.
29. Geenen DL, Gilliam TB, Crowley D, Moorehead-Steffens C, Rosenthal A. Echocardiographic measures in 6 to 7 year old children after an 8 month exercise program. *Am J Cardiol.* 1982;49(8):1990–1995.
30. Obert P, Mandigout S, Vinet A, N'Guyen LD, Stecken F, Courteix D. Effect of aerobic training and detraining on left ventricular dimensions and diastolic function in prepubertal boys and girls. *Int J Sports Med.* 2001;22(2):90–96.
31. Wu WC, Chang LY, Luh DL, et al. Sex differences in the trajectories of and factors related to extracurricular sport participation and exercise: A cohort study spanning 13 years. *BMC Public Health.* 2020;20(1):1639.
32. Diaz-Canestro C, Siebenmann C, Montero D. Blood oxygen carrying capacity determines cardiorespiratory fitness in middle-age and older women and men. *Med Sci Sports Exerc.* 2021;53(11):2274–2282.
33. Murphy WG. The sex difference in haemoglobin levels in adults—mechanisms, causes, and consequences. *Blood Rev.* 2014;28(2):41–47.
34. Morkeberg J, Lundby C, Nissen-Lie G, Nielsen TK, Hemmersbach P, Damsgaard R. Detection of darbepoetin alfa misuse in urine and blood: A preliminary investigation. *Med Sci Sports Exerc.* 2007;39(10):1742–1747.

35. Lundby C, Montero D. Did you know-why does maximal oxygen uptake increase in humans following endurance exercise training? *Acta Physiol (Oxf)*. 2019;227(4):e13371.
36. Jacobs RA, Lundby C, Robach P, Gassmann M. Red blood cell volume and the capacity for exercise at moderate to high altitude. *Sports Med*. 2012;42(8):643–663.
37. Lundby C, Thomsen JJ, Boushel R, et al. Erythropoietin treatment elevates haemoglobin concentration by increasing red cell volume and depressing plasma volume. *J Physiol*. 2007;578(Pt 1):309–314.
38. Rasmussen P, Siebenmann C, Diaz V, Lundby C. Red cell volume expansion at altitude: A meta-analysis and Monte Carlo simulation. *Med Sci Sports Exerc*. 2013;45(9):1767–1772.
39. Bonne TC, Lundby C, Jorgensen S, et al. 'Live High-Train High' increases hemoglobin mass in Olympic swimmers. *Eur J Appl Physiol*. 2014;114(7):1439–1449.
40. Gore CJ, Sharpe K, Garvican-Lewis LA, et al. Altitude training and haemoglobin mass from the optimised carbon monoxide rebreathing method determined by a meta-analysis. *Br J Sports Med*. 2013;47 Suppl 1:i31–i39.
41. Schmidt W, Prommer N. Effects of various training modalities on blood volume. *Scand J Med Sci Sports*. 2008;18 Suppl 1:57–69.
42. Lundby C, Robach P. Does 'altitude training' increase exercise performance in elite athletes? *Exp Physiol*. 2016;101(7):783–788.
43. Klein M, Kaestner L, Bogdanova AY, et al. Absence of neocytolysis in humans returning from a 3-week high-altitude sojourn. *Acta Physiol (Oxf)*. 2021;232(3):e13647.
44. Siebenmann C, Dempsey JA. Hypoxic training is not beneficial in elite athletes. *Med Sci Sports Exerc*. 2020;52(2):519–522.
45. Ploszczyca K, Langfort J, Czuba M. The effects of altitude training on erythropoietic response and hematological variables in adult athletes: A narrative review. *Front Physiol*. 2018;9:375.
46. Azad P, Villafuerte FC, Bermudez D, Patel G, Haddad GG. Protective role of estrogen against excessive erythrocytosis in Monge's disease. *Exp Mol Med*. 2021;53(1):125–135.
47. Calbet JA, Gonzalez-Alonso J, Helge JW, et al. Central and peripheral hemodynamics in exercising humans: Leg vs arm exercise. *Scand J Med Sci Sports*. 2015;25 Suppl 4:144–157.
48. Secher NH, Ruberg-Larsen N, Binkhorst RA, Bonde-Petersen F. Maximal oxygen uptake during arm cranking and combined arm plus leg exercise. *J Appl Physiol*. 1974;36(5):515–518.
49. Diaz-Canestro C, Pentz B, Sehgal A, Montero D. Sex dimorphism in cardiac and aerobic capacities: The influence of body composition. *Obesity (Silver Spring)*. 2021;29(11):1749–1759.

50. Calbet JA, Lundby C, Sander M, Robach P, Saltin B, Boushel R. Effects of ATP-induced leg vasodilation on VO2 peak and leg O2 extraction during maximal exercise in humans. *Am J Physiol Regul Integr Comp Physiol.* 2006;291(2):R447–R453.
51. Bada AA, Svendsen JH, Secher NH, Saltin B, Mortensen SP. Peripheral vasodilatation determines cardiac output in exercising humans: Insight from atrial pacing. *J Physiol.* 2012;590(8):2051–2060.
52. Gonzalez-Alonso J, Mortensen SP, Jeppesen TD, et al. Haemodynamic responses to exercise, ATP infusion and thigh compression in humans: Insight into the role of muscle mechanisms on cardiovascular function. *J Physiol.* 2008;586(9):2405–2417.
53. Robbins JL, Duscha BD, Bensimhon DR, et al. A sex-specific relationship between capillary density and anaerobic threshold. *J Appl Physiol.* 2009;106(4):1181–1186.
54. Diaz-Canestro C, Pentz B, Sehgal A, Montero D. Sex differences in cardiorespiratory fitness are explained by blood volume and oxygen carrying capacity. *Cardiovasc Res.* 2021;118(1):334–343.
55. Diaz-Canestro C, Pentz B, Sehgal A, Montero D. Differences in cardiac output and aerobic capacity between sexes are explained by blood volume and oxygen carrying capacity. *Front Physiol.* 2022;13:747903.
56. Janssen I, Heymsfield SB, Wang ZM, Ross R. Skeletal muscle mass and distribution in 468 men and women aged 18–88 yr. *J Appl Physiol.* 2000;89(1):81–88.
57. Calbet JA, Joyner MJ. Disparity in regional and systemic circulatory capacities: Do they affect the regulation of the circulation? *Acta Physiol (Oxf).* 2010;199(4):393–406.
58. Guo M, Diaz Canestro C, Pugliese NR, Paneni F, Montero D. Lean body mass and the cardiovascular system: An intrinsic relationship in Hans Chinese women and men. Hong Kong College of Cardiology Annual Scientific Congress (HKCC ASC 2023); 2023; Hong Kong.
59. Wilmore JH, Parr RB, Girandola RN, et al. Physiological alterations consequent to circuit weight training. *Med Sci Sports.* 1978;10(2):79–84.
60. Francis G. https://science4performance.com/2019/09/12/cycling-physique/. 2019. Accessed 25.12.2020.
61. Haakonssen EC, Barras M, Burke LM, Jenkins DG, Martin DT. Body composition of female road and track endurance cyclists: Normative values and typical changes. *Eur J Sport Sci.* 2016;16(6):645–653.

Epilogue

What ought to be said to *close* a book about physiology? First and foremost, our field of study is far from allowing closure. If someone, making the parallelism with hard sciences (e.g., physics, chemistry), claims that the laws of physiology are 'carved in stone', it must be on jelly-like matter. In fact, current knowledge regarding the functional attributes of the human body is, to a certain degree, inherently fragmentary and arbitrary. A number of facts are commonly selected, which compound a rather coherent narrative, one that sells textbooks and scientific publications. It is not merely an oblivious attitude, a self-imposed blindfoldedness towards seemingly contradictory evidence. It is a deeper issue. The inveterate tradition of partitioning the functions of the body among 'distinct' anatomical systems already imposes major implicit limitations. For instance, how many relevant knowledge gaps and misinterpretations would be prevented if we were to *see* and therefore teach cardiovascular, lymphatic and renal systems as a single functional unit responsible, in an integrative manner, for the volume and flow of fluid throughout the body? The small subfield of integrative physiology is an attempt, a largely incomplete one, to be closer to the facts—all of them. Unlike molecular or cell biology, physiology should be, as a matter of principle, as integrative as possible.

To fully grasp how the human body works, we need to understand the messily intertwined products of evolution. Such an unconscious process is not urged to generate neatly ordered and understandable machines, as we humans do to facilitate our existence. It is indeed on this point that the human body and human-made machines essentially differ. We will never take as an example of human-made machine anything being the product of idle design, we will instead

DOI: 10.1201/9781003486893-7

call it an accident. This is precisely the most rational definition for biological products. Parenthetically, it is long overdue that we state it clearly: there is no purpose anywhere, other than in our imagination. Our body is the result of cumulative accidents that reproduce themselves with eventual errors (mutations) from at least 3.5 billion years ago. What could be expected from a very long time of random minute mutations of extremely simple biomolecules that progressively accrued complexity, the whole process being winnowed by reproductive success and never starting anew to correct long-term *strategical* blunders? Our physiology, and that of any living being in the universe, is expected, logically, to partly comprise senselessness 'designs' and messily intertwined mechanisms.

A glaringly obvious example of absurd structure is epitomized by the human retina—a camera-like system with the light sensors installed backwards relative to the beam of light and the 'wiring' (axons) in-between to cap it all. Besides the obvious functional issues derived from such an arrangement, it contributes to the occurrence of retinal detachment in humans. And the retina is not the only *oversight*. To find another, we do not need to look far. Have you ever wondered why humans are so prone to get colds and sinus infections? The drainage hole of the largest pair of mucus-storing cavities, i.e., the maxillary sinuses, is located at the top rather than the bottom of these anatomical structures. This foolish placement hampers mucus drainage (thereby fostering infections) due to this annoying force called gravity. A few centimeters below, in the neck, again we encounter a striking arrangement. The recurrent laryngeal nerve, which in any reasonably efficient design should directly travel from the brain to the larynx, instead makes an outrageous detour through the heart's main large vessels in the chest. A pointless route, other than to test the skills of cardiac surgeons. Yet, collectively considered, it would be unfair to generally characterize our body as the product of foolish design. Likewise, it would be small-minded to marvel at the organs and systems that do work apparently fine while neglecting what does not fit in a nice textbook story.

Accidents, coincidences and randomness are concepts thoroughly comprehended and never forgotten by some physicians—those who frequently face critical situations are constrained to not digress from the fundaments of reality. Vastly distinct perspectives on the same issue arise when the external check is *truth* rather than the opinion of academic peers. It is in this context that the typical disdain of physicians for physiology, as an academic endeavor, can be sensibly understood. Reality can seldom be fairly described with well-rounded and cohesive stories. Not to mention the overwhelming constraints and necessary shortcomings of modern academics to develop honest work.[1, 2]

Notwithstanding the myriad of elementary and circumstantial limitations that humans need to overcome or evade to decipher the fundaments of physiology, we know something. Through a very slow tortuous path, we have conquered a certain body of knowledge. In fact, we can largely modify exercise capacity in women and men via the quantitative manipulation of a single protein in our blood. Eventually, we might know virtually all that is relevant in our body, and looking back, we will be surprised at the haphazard principles that govern us—such as our amazement nowadays, via the power of computers, with the absence of *rational* principles in chess. At the same time, our future functional capacities will prevalently entrain human-made artifacts, i.e., *rational* elements (e.g., cardiac pacemakers, joint replacements, blood glucose monitors), whose principles must be fully understandable and straightforward to manipulate. Yet, the power to modify key physiological functions ad libitum, without negative collateral effects, will come through the integrative understanding of the human body, leading to the effective discernment of predominant, ancillary and incidental mechanisms. In this book, no attempt was made to be merely comprehensive, which is indeed currently unrealistic. Instead, I took a stance, putting the focus on crucial mechanisms according to my judgement of, and sometimes detailed experience with, the available devidence. I hope the reader was ignited with a kind of skepticism that craves for truer knowledge, never assimilated at face value.

REFERENCES

1. Almog T, Almog O. *Academia: All the Lies: What Went Wrong in the University Model and What Will Come in Its Place*. Independently published;2020.
2. Broad W, Wade N. *Betrayers of the Truth: Fraud and Deceit in the Halls of Science*. New York: Simon & Schuster;1983.

Index

Page numbers in *italic* indicate figures.

Printed in the United States
by Baker & Taylor Publisher Services